Geometry

Written by
Pamela Jennett

Editor: Collene Dobelmann
Illustrator: Carmela Murray
Designer/Production: Moonhee Pak/Carmela Murray
Cover Designer: Barbara Peterson
Art Director: Tom Cochrane
Project Director: Carolea Williams

© 2004 Creative Teaching Press, Inc., Huntington Beach, CA 92649
Reproduction of activities in any manner for use in the classroom and not for commercial sale is permissible.
Reproduction of these materials for an entire school or for a school system is strictly prohibited.

Table of Contents

Introduction

Each book in the *Power Practice*™ series contains dozens of ready-to-use activity pages to provide students with skill practice. The fun activities can be used to supplement and enhance what you are already teaching in your classroom. Give an activity page to students as independent class work, or send the pages home as homework to reinforce skills taught in class. An answer key is included at the end of each book to provide verification of student responses.

Geometry provides students with an introduction to and practice of the basic concepts of plane geometry. The topics covered in this resource include lines and segments, plane figures, angles, measurement, and transformations. Students also have opportunities to use algebraic equations to solve variables present in geometric figures. The activities provide excellent supplementation for textbooks or any comprehensive study of geometry.

Use these ready-to-go activities to "recharge" skill review and give students the power to succeed!

Geometric Vocabulary

Geometry has its own vocabulary to describe certain types of geometric objects. Match each term in the word box to its definition. Use the same terms to label the illustrations.

> ray   angle   line   line segment
> plane   point   collinear   coplanar

1 _____ This is a geometric object with no dimensions. It is only a location.

2 _____ This is a collection of points along a straight path. It has no endpoints.

3 _____ This describes points that lie on the same line.

4 _____ This geometric object is a flat surface that extends endlessly in all directions.

5 _____ This describes points that lie on the same plane.

6 _____ This is a part of a line that has two endpoints.

7 _____ This is a part of a line that has only one endpoint.

8 _____ This geometric object is formed from two rays with a common endpoint.

9

10

11

12

13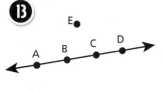

14

15

16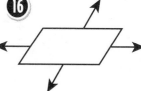

Geometry © 2004 Creative Teaching Press

Identify Lines, Line Segments, and Rays

Points, lines, and planes are the building blocks of geometry.

$\overleftrightarrow{XY}$ $\overleftrightarrow{YZ}$ $\overleftrightarrow{XZ}$

These name **lines** in the diagram. Lines have no endpoint, but continue on in both directions.

$\overrightarrow{AB}$ $\overrightarrow{AC}$ $\overrightarrow{BC}$

These name **rays** in the diagram. Rays have an endpoint and continue in a single direction.

$\overline{DE}$ $\overline{ED}$

These name **line segments**. A line segment has two endpoints.

Complete.

1 Give three names for this line.

_____ _____ _____

2 Name three rays.

_____ _____ _____

3 Name three line segments.

_____ _____ _____

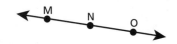

4 Give three names for this line.

_____ _____ _____

5 Name three rays.

_____ _____ _____

6 Name three line segments.

_____ _____ _____

Geometry © 2004 Creative Teaching Press

Name Lines, Line Segments, and Rays

symbol: $\longleftrightarrow$

To name a line, give two points along that line. Place the symbol for line over the top. The points can be written in either order.

$\overleftrightarrow{MO}$ $\overleftrightarrow{OM}$

symbol: $\longrightarrow$

To name a ray, give the starting point first, then a following point on the ray. The name can only be written in the direction the ray travels.

$\overrightarrow{RT}$ $\overrightarrow{RX}$

symbol: $\longrightarrow$

To name a segment, give two points that mark a segment. The points can be written in either order.

$\overline{AB}$ $\overline{BA}$

Circle the correct ways to name each of the following.

1

This is a ray.

$\overrightarrow{JK}$ $\overrightarrow{KL}$ $\overrightarrow{LK}$

$\overline{LJ}$ $\overleftrightarrow{JK}$ $\overrightarrow{JL}$

2

This is a line.

$\overleftrightarrow{RU}$ $\overleftrightarrow{UV}$ $\overleftrightarrow{RV}$

$\overleftrightarrow{VU}$ $\overrightarrow{UV}$ $\overline{RV}$

3

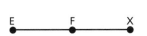

This is a line segment.

$\overline{EF}$ $\overleftrightarrow{FX}$ $\overline{XF}$

$\overline{EX}$ $\overleftrightarrow{XF}$ $\overline{FE}$

Geometry © 2004 Creating Teaching Press

Find Lengths of Line Segments

Because a line segment has two endpoints, it has a definite length.

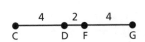

The length of $\overline{CD}$ is 4 units. Write the length as $\overline{CD} = 4$.

Two line segments that have the same length are congruent.

In this diagram $\overline{FG}$ is congruent to $\overline{CD}$.

Write this congruency as $\overline{FG} \cong \overline{CD}$.

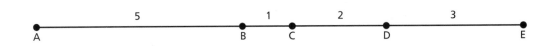

Find the length of these line segments.

1 $\overline{AB}$ = _____

2 $\overline{BC}$ = _____

3 $\overline{CD}$ = _____

4 $\overline{DE}$ = _____

5 $\overline{AC}$ = _____

6 $\overline{DA}$ = _____

7 $\overline{EA}$ = _____

8 $\overline{DB}$ = _____

9 $\overline{AD}$ = _____

10 $\overline{BE}$ = _____

11 $\overline{CE}$ = _____

12 $\overline{AE}$ = _____

Name line segments that are congruent to the following:

13 $\overline{AB} \cong$ _____

14 $\overline{DE} \cong$ _____

15 $\overline{EC} \cong$ _____

16 $\overline{BD} \cong$ _____

Geometry © 2004 Creative Teaching Press

Collinear Points and Line Segments

Collinear points are points that lie on the same line.

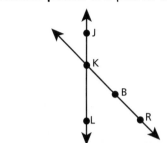

Points J and L are on the same line.
They are collinear.

Points J and B are not on the same line.
They are not collinear.

$\overline{JK}$ and $\overline{LK}$ are collinear. $\overline{RB}$ and $\overline{JK}$ are not.

Read each statement. Use the diagram and write if the statement is **true** or **false**.

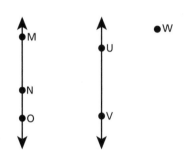

1 Point M and Point U are collinear. _____

2 Point N and Point O are collinear. _____

3 $\overline{MN}$ and $\overline{UV}$ are not collinear. _____

4 $\overline{UV}$ and $\overline{ON}$ are collinear. _____

5 Point W and Point V are collinear. _____

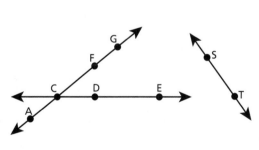

6 Point C and Point G are collinear. _____

7 $\overline{AC}$ and $\overline{DE}$ are collinear. _____

8 $\overline{GF}$ and $\overline{ST}$ are not collinear. _____

9 Point E and Point S are collinear. _____

10 Point F and Point A are collinear. _____

11 Point C and Point E are not collinear. _____

12 $\overline{AC}$ and $\overline{GF}$ are collinear. _____

Geometry © 2004 Creative Teaching Press

Name _____ Date _____

Identify Objects in a Plane

Each face of a cube lies on a different **plane**. A plane is a flat surface extending indefinitely in all directions. One of these planes is represented in the diagram.

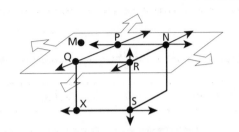

Point M and Point N are on this plane. They are coplanar.

Point S does not lie on the same plane.

$\overleftrightarrow{PQ}$ and $\overleftrightarrow{RN}$ lie on the same plane. They are coplanar.

$\overleftrightarrow{PQ}$ and $\overleftrightarrow{RS}$ are not on the same plane.

Use the diagram above to solve.

1 Name two points that are coplanar to Point P. _____

2 Name a point that is not coplanar to Point P. _____

3 Name a line that is coplanar to $\overleftrightarrow{RN}$. _____

4 Name a line that is not coplanar to $\overleftrightarrow{RN}$. _____

5 Name all the points shown that are coplanar to Point M.

_____

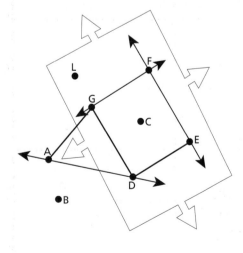

6 Name a point that is not coplanar to Point L.

_____

7 Name a line that is coplanar to $\overleftrightarrow{FE}$. _____

8 Name a line that is not coplanar to $\overleftrightarrow{FE}$. _____

9 Name two points that are coplanar to Point C.

_____

10 Name two line segments that are coplanar to $\overline{ED}$.

_____

Geometry © 2004 Creative Teaching Press

Geometry © 2004 Creative Teaching Press

Name _____ Date _____

Line and Segment Relationships

Lines and line segments can be described by their relationships to each other in several ways. Use the terms in the word box to label each relationship. Then write the symbol or words used to show the relationship mathematically.

| skew | concurrent | intersecting | parallel | perpendicular |

	Relationship	Example	Definition	Write
1			Lines that meet at a point.	
2			Three or more lines that meet at the same point.	
3			Lines that intersect to form a right angle.	
4			Lines in the same plane that do not intersect.	
5			Lines not in the same plane that do not intersect.	

Intersecting and Parallel Lines in Planes

The intersection of two or more geometric figures is the set of points that the figures have in common.

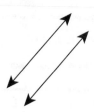

Point C

These two lines **intersect** at Point C.

Two lines are parallel if they are on the same plane and do not intersect.

These two lines are **parallel.** They do not intersect at any point, even if the lines continued on indefinitely.

Use the diagram to solve.

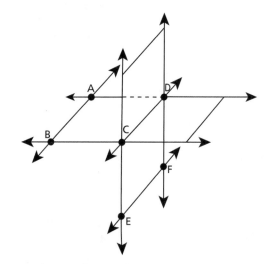

1 Which lines intersect $\overleftrightarrow{CD}$?

_____

2 Which lines are parallel to $\overleftrightarrow{CD}$?

_____

3 Which lines intersect $\overleftrightarrow{FE}$?

_____

4 Which lines are parallel to $\overleftrightarrow{FE}$?

_____

Write **true** or **false** for each statement. If false, explain why.

5 $\overleftrightarrow{AB}$ intersects $\overleftrightarrow{DC}$. _____

6 $\overleftrightarrow{EF}$ is parallel to $\overleftrightarrow{CD}$. _____

7 $\overleftrightarrow{BA}$ intersects $\overleftrightarrow{CD}$. _____

8 $\overleftrightarrow{EF}$ is parallel to $\overleftrightarrow{BA}$. _____

Geometry © 2004 Creative Teaching Press

Name _____ Date _____

Identify Intersecting and Parallel Segments

Line segments intersect if they share a common point.

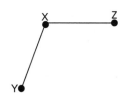

These line segments intersect at Point X.

Line segments are parallel if they are on the same plane and do not intersect.

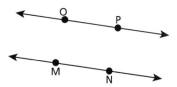

These line segments are parallel. They are found on lines that do not intersect.

Use the diagram to solve. Write **true** or **false** for each statement.

1 $\overline{HG}$ intersects $\overline{IB}$. _____

2 $\overline{HG}$ intersects $\overline{HD}$. _____

3 $\overline{EH}$ intersects $\overline{HC}$. _____

4 $\overline{CH}$ intersects $\overline{IB}$. _____

5 $\overline{IF}$ intersects $\overline{AI}$. _____

6 $\overline{GH}$ is parallel to $\overline{DH}$. _____

7 $\overline{CH}$ is parallel to $\overline{IA}$. _____

8 $\overline{HD}$ is parallel to $\overline{IB}$. _____

9 $\overline{IB}$ is parallel to $\overline{EH}$. _____

10 $\overline{FI}$ is parallel to $\overline{GH}$. _____

Geometry © 2004 Creative Teaching Press

Find Geometric Answers

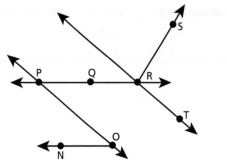

These figures are in the same plane. Use the diagram to solve.

1 Write four other names for $\overleftrightarrow{PQ}$.

_____

_____

2 Name three different lines.

_____

_____

3 Name a parallel line to $\overleftrightarrow{PO}$.

_____

_____

4 Name a line that intersects $\overrightarrow{RS}$.

_____

_____

5 Name three lines that do not intersect.

_____

_____

6 Name two pairs of intersecting lines.

_____

_____

7 Is ON a ray or a line? Explain.

_____

_____

8 Are $\overleftrightarrow{PQ}$ and $\overleftrightarrow{RQ}$ the same line? Explain.

_____

_____

9 Is $\overleftrightarrow{XY}$ parallel to $\overleftrightarrow{PO}$? Explain.

_____

_____

10 Are $\overleftrightarrow{XY}$ and $\overrightarrow{RS}$ intersecting? Explain.

_____

_____

Geometry © 2004 Creative Teaching Press

Name _____ Date _____

Congruence of Segments

Congruent segments have the same length. The symbol for congruent is ≅.

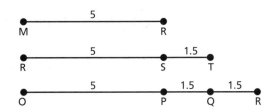

Congruent segments:

$\overline{AB}$ = 4 and $\overline{CD}$ = 4 $\overline{AB} \cong \overline{CD}$

$\overline{AB}$ = 4 and $\overline{BC}$ = 2 $\overline{AB}$ and $\overline{BC}$ are not congruent.

Use the diagram. Write **true** or **false** for each statement.

❶ $\overline{MR} \cong \overline{RS}$ _____

❷ $\overline{ST} \cong \overline{QP}$ _____

❸ $\overline{OP} \cong \overline{TS}$ _____

❹ $\overline{PQ} \cong \overline{RM}$ _____

❺ $\overline{QP} \cong \overline{ST}$ _____

❻ $\overline{SR} \cong \overline{MR}$ _____

❼ $\overline{RS} \cong \overline{OP}$ _____

❽ $\overline{QR} \cong \overline{ST}$ _____

Find the length of the indicated segments. Circle the congruent segments in each row.

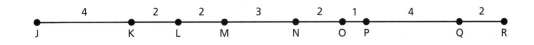

❾ $\overline{JK}$ = _____ $\overline{LM}$ = _____ $\overline{MN}$ = _____ $\overline{PQ}$ = _____

❿ $\overline{NO}$ = _____ $\overline{MN}$ = _____ $\overline{QR}$ = _____ $\overline{KL}$ = _____

⓫ $\overline{KL}$ = _____ $\overline{OP}$ = _____ $\overline{QR}$ = _____ $\overline{PQ}$ = _____

⓬ $\overline{MN}$ = _____ $\overline{ON}$ = _____ $\overline{QP}$ = _____ $\overline{KJ}$ = _____

Geometry © 2004 Creative Teaching Press

Add to Find Congruent Segments

Add segments to find the length of a new segment.

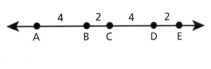

Is $\overline{AC} \cong \overline{CE}$?

1. Find lengths of given segments. Add.

$\overline{AB} + \overline{BC} = \overline{AC}$

$4 + 2 = 6 \qquad \overline{AC} = 6$

$\overline{CD} + \overline{DE} = \overline{CE}$

$4 + 2 = 6 \qquad \overline{AC} = 6$

2. Compare given segments.

$\overline{AC} \cong \overline{CE}$

Use the diagram. Write **true** or **false** for each statement.

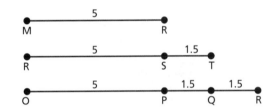

1 $\overline{MR} \cong \overline{RT}$ _____

2 $\overline{MR} \cong \overline{RP}$ _____

3 $\overline{OP} \cong \overline{TR}$ _____

4 $\overline{PR} \cong \overline{RS}$ _____

5 $\overline{QO} \cong \overline{RT}$ _____

6 $\overline{SR} \cong \overline{MR}$ _____

7 $\overline{RT} \cong \overline{QO}$ _____

8 $\overline{RP} \cong \overline{RT}$ _____

Find the length of the indicated segments. Circle the congruent segments in each row.

9 $\overline{JK}$ = _____ $\overline{JM}$ = _____ $\overline{MK}$ = _____ $\overline{RP}$ = _____

10 $\overline{LO}$ = _____ $\overline{MN}$ = _____ $\overline{PR}$ = _____ $\overline{KN}$ = _____

11 $\overline{MJ}$ = _____ $\overline{OL}$ = _____ $\overline{QN}$ = _____ $\overline{PR}$ = _____

12 $\overline{MP}$ = _____ $\overline{JM}$ = _____ $\overline{KN}$ = _____ $\overline{LP}$ = _____

Geometry © 2004 Creative Teaching Press

Subtract to Find Segment Lengths

Use what is known about line segment lengths. Subtract to find missing line segment measurements.

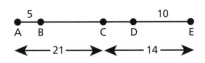

1. Use the lengths you know.
   $\overline{AC} = \overline{AB} + \overline{BC}$ $\overline{AC} = 21$ $\overline{AB} = 5$

2. Subtract given lengths.
   $21 - 5 = 16$

3. $\overline{BC} = 16$

Use each diagram to find the lengths of the given segments.

1 $\overline{CE} =$ _____ $\overline{DE} =$ _____ $\overline{CF} =$ _____

2 $\overline{QR} =$ _____ $\overline{PS} =$ _____ $\overline{ST} =$ _____

3 $\overline{JK} =$ _____ $\overline{JM} =$ _____ $\overline{KM} =$ _____

4 $\overline{SV} =$ _____ $\overline{UV} =$ _____ $\overline{TW} =$ _____

Use the diagram to find the segment lengths.

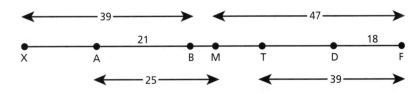

5 $\overline{XA} =$ _____   **6** $\overline{AB} =$ _____   **7** $\overline{BM} =$ _____

8 $\overline{MT} =$ _____   **9** $\overline{TD} =$ _____   **10** $\overline{DF} =$ _____

11 $\overline{XB} =$ _____   **12** $\overline{BT} =$ _____   **13** $\overline{AT} =$ _____

Geometry © 2004 Creative Teaching Press

Find Congruent Line Segments

Use what is known to find the missing segment lengths.

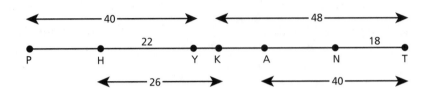

Find the length of these line segments.

1 $\overline{PH}$ = _____   **2** $\overline{PY}$ = _____

3 $\overline{PK}$ = _____   **4** $\overline{PA}$ = _____

5 $\overline{YK}$ = _____   **6** $\overline{YA}$ = _____

7 $\overline{YN}$ = _____   **8** $\overline{YT}$ = _____

9 $\overline{HK}$ = _____   **10** $\overline{HA}$ = _____

11 $\overline{HT}$ = _____   **12** $\overline{KA}$ = _____

13 $\overline{KN}$ = _____   **14** $\overline{KT}$ = _____

15 $\overline{AN}$ = _____   **16** $\overline{AT}$ = _____

Use the diagram and your information above. Are these segments congruent? Write **true** or **false**.

17 $\overline{PH} \cong \overline{TN}$ _____   **18** $\overline{HK} \cong \overline{YA}$ _____

19 $\overline{PY} \cong \overline{YN}$ _____   **20** $\overline{TA} \cong \overline{YP}$ _____

21 $\overline{HK} \cong \overline{YA}$ _____   **22** $\overline{HY} \cong \overline{NA}$ _____

23 $\overline{HK} \cong \overline{KN}$ _____   **24** $\overline{PK} \cong \overline{KT}$ _____

25 $\overline{YP} \cong \overline{YN}$ _____   **26** $\overline{PY} \cong \overline{AT}$ _____

Geometry © 2004 Creative Teaching Press

Geometry and Algebraic Equations

In order to solve for missing line segment lengths, use the given information. Write an algebraic equation and solve for *x*.

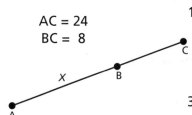

AC = 24
BC = 8

1. Find the value of the segment marked *x*.

2. The length of $\overline{AB}$ and $\overline{BC}$ is equal to the length of $\overline{AC}$.
   AB + BC = AC

3. Replace with what you know.
   x + 8 = 24

4. Solve for *x*.
   24 − 8 = 16   *x* = 16

For each diagram, use an algebraic equation and solve for *x*.

1 LP = 31
       LM = 12

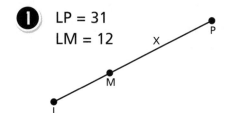

LM + *x* = LP

_____

_____

2 TU = 27

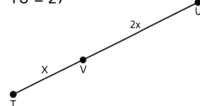

x + 2*x* = TU

_____

_____

3

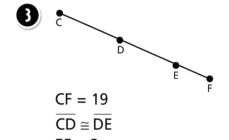

CF = 19
$\overline{CD} \cong \overline{DE}$
EF = 3

x + *x* + 3 = CF

_____

_____

4

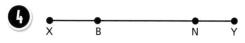

$\overline{XB} \cong \overline{NY}$
XY = 9
XB = *x* − 3

(*x* − 3) + *x* + (*x* − 3) = 9

_____

_____

Angle Vocabulary

Geometry uses special vocabulary to describe and classify angles and their parts. Match each term in the word box to its definition.

vertical	adjacent	vertex	side	linear	acute
obtuse	right	straight	complementary	supplementary	bisector

① _____ This is the common endpoint of two rays that form an angle.

② _____ This describes a pair of adjacent angles whose noncommon sides are opposite rays.

③ _____ This describes any two non-overlapping angles that share a common ray and a common vertex.

④ _____ This describes an angle having a measure greater than 90° and less than 180°.

⑤ _____ This refers to one of the rays that form an angle.

⑥ _____ This describes an angle that has a measure of exactly 90°.

⑦ _____ These are also called opposite angles. They are nonadjacent angles formed by intersecting lines.

⑧ _____ This describes a pair of angles with combined measures that equal 90°.

⑨ _____ This describes a pair of angles with combined measures that equal 180°.

⑩ _____ This describes an angle that measures exactly 180°.

⑪ _____ This refers to a point, line, or plane that divides a geometric figure into congruent halves.

⑫ _____ This describes an angle having a measure greater than 0° and less than 90°.

Geometry © 2004 Creative Teaching Press

Name _____ Date _____

Name and Label Angles

Angles are formed when two rays share a common endpoint.

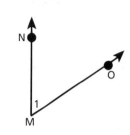

Point M is the vertex.
$\overrightarrow{MN}$ and $\overrightarrow{MO}$ are the sides.

There are four ways to name this angle:
∠1 , ∠M, ∠NMO, or ∠OMN

When naming angles by their points, the vertex point must always be the center letter.

Use the diagram to complete the information.

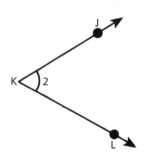

1 Vertex: _____

2 Sides: _____ and _____

3 Four names for this angle: _____

_____

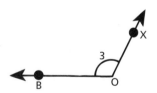

4 Vertex: _____

5 Sides: _____ and _____

6 Four names for this angle: _____

_____

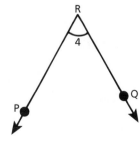

7 Vertex: _____

8 Sides: _____ and _____

9 Four names for this angle: _____

_____

Geometry © 2004 Creative Teaching Press

Name and Label Connected Angles

Angles can be connected to other angles.

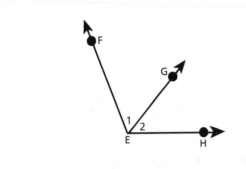

Point E is the vertex for both angles.
$\overrightarrow{EG}$ is the common side.

One angle is ∠1 .
∠1 can be named ∠FEG or ∠GEF.

The second angle is ∠2.
∠2 can be named ∠GEH or ∠HEG.

The last angle can be named ∠FEH or ∠HEF.
None can be called ∠E because there is more than one angle with that vertex.

Use the diagram to complete the information.

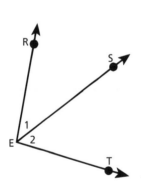

1 Common vertex: _____

2 Common side: _____

3 Names for first angle: _____

_____

4 Names for second angle: _____

_____

5 Names for third angle: _____

_____

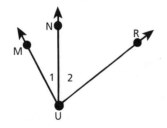

6 Common vertex: _____

7 Common side: _____

8 Names for three different angles: _____

_____

Geometry © 2004 Creative Teaching Press

Name _____ Date _____

Classify Angles

Angles are measured in units called **degrees**. An angle can be classified by its measure.

Acute  **Right** **Obtuse** **Straight**

| Acute angles have measures greater than 0° and less than 90°. | Right angles have measures equal to 90°. Notice the symbol for right angle. | Obtuse angles have measures greater than 90° and less than 180°. | Straight angles have measures equal to 180°. |

Use the types of angles above to label each illustration. Write three names for each angle.

1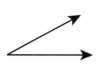

Angle: _____

_____

_____

2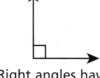

Angle: _____

_____

_____

3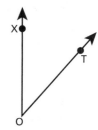

Angle: _____

_____

_____

4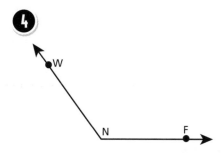

Angle: _____

_____

_____

5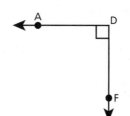

Angle: _____

_____

_____

6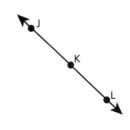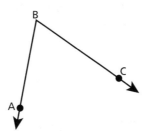

Angle: _____

_____

_____

Geometry © 2004 Creative Teaching Press

Name _____ Date _____

Identify Congruent Angles

Angles with the same measure are congruent.

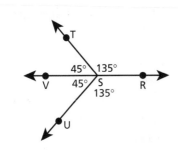

In this diagram, $\overleftrightarrow{VR}$ bisects the right angle TSU to form two sets of congruent angles.

∠TSV and ∠USV are acute angles with the same measure. They are congruent.
∠TSV ≅ ∠USV

∠TSR and ∠USR are obtuse angles with the same measure.
∠TSR ≅ ∠USR

Use the diagram to find congruent angles.

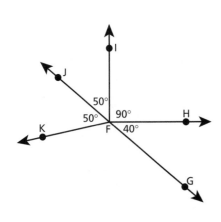

1 Name three acute angles. _____

2 Name two obtuse angles. _____

3 Name one right angle. _____

4 Which angles are congruent? _____

_____

5 Name a bisector. _____

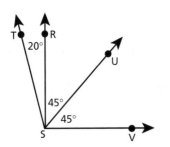

6 Name four acute angles. _____

7 Name one obtuse angle. _____

8 Name one right angle. _____

9 Which angles are congruent? _____

_____

10 Name a bisector. _____

Geometry © 2004 Creative Teaching Press

Add and Subtract to Find Angle Measures

Find the measure of angles by adding and subtracting known measures.

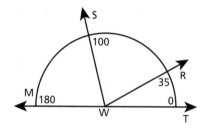

Find the measure of ∠SWR.

This diagram shows the measure of each angle from 0°. Measure can be abbreviated as *m*.

m∠RWT = 35°   m∠SWT = 100°

To find m∠SWR, subtract the known measures.
m∠SWT − m∠RWT = m∠SWR.
100° − 35° = 65°
m∠SWR = 65°

Use the diagram. Show how you add or subtract known measures to find the measures of the given angles.

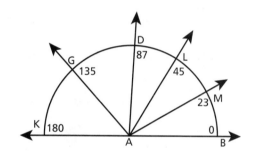

1 m∠MAB = _____

2 m∠LAB = _____

3 m∠DAB = _____

4 m∠GAB = _____

5 m∠KAB = _____

6 m∠DAM = _____

7 m∠KAG = _____

8 m∠DAK = _____

9 m∠LAG = _____

10 m∠MAG = _____

11 m∠LAK = _____

12 Name two pairs of congruent angles.

_____ ≅ _____

_____ ≅ _____

Geometry © 2004 Creative Teaching Press

Calculate Angle Measures

Label the diagram using the given measures. Then find the measure of angles by adding and subtracting known measures.

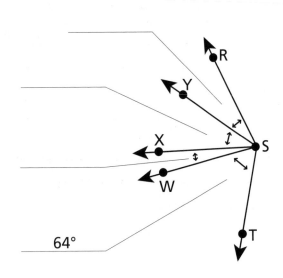

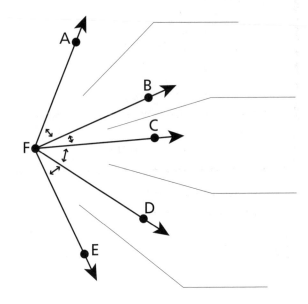

Label the angles. The first one is done for you.

1 m∠WST = 64°

2 m∠RSY = 30°

3 m∠WSX = 11°

4 m∠XSY = 33°

Find these measures:

5 m∠RSX = _____

6 m∠RST = _____

7 m∠TSX = _____

8 m∠WSY = _____

9 m∠TSY = _____

10 m∠WSR = _____

Label the angles.

11 m∠AFB = 45°

12 m∠BFC = 22°

13 m∠DFE = 33°

14 m∠CFD = 41°

Find these measures:

15 m∠AFC = _____

16 m∠DFB = _____

17 m∠AFD = _____

18 m∠BFE = _____

19 m∠EFA = _____

20 m∠EFC = _____

Geometry © 2004 Creative Teaching Press

Complementary and Perpendicular Angles

Two angles are **complementary** if the sum of their measures is 90°. Each angle is the complement of the other.

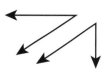

Adjacent complementary angles share a common side and vertex. The sum of the two angle measures is 90°.

Nonadjacent complementary angles do not share a common side or vertex. The sum of the two angle measures is 90°.

Use the given measure to determine the complementary angle's measure. Label each illustration as **adjacent** or **nonadjacent**.

1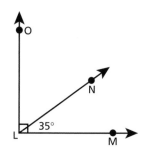

m∠OLN = _____

m∠NLM = _____

Type: _____

2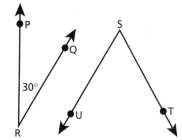

m∠PRQ = _____

m∠UST = _____

Type: _____

3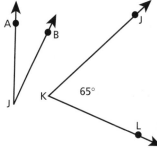

m∠JKL = _____

m∠AJB = _____

Type: _____

4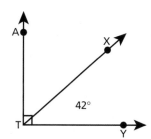

m∠XTY = _____

m∠ATX = _____

Type: _____

5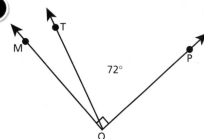

m∠TOP = _____

m∠MOT = _____

Type: _____

6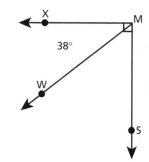

m∠XMW = _____

m∠SMW = _____

Type: _____

Geometry © 2004 Creative Teaching Press

Supplementary and Linear Angles

Two angles are supplementary if the sum of their measures is 180°. Each angle is the supplement of the other. If the supplementary angles are adjacent, then they are also known as a linear pair.

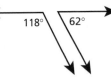

Supplementary Pair

Linear Pair

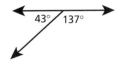

The sum of the angles is 180°. They are not adjacent so they are a supplementary pair.

The sum of these adjacent angles is 180°. If the two sides they don't share form opposite rays, then they are also called a linear pair.

Use the given measure to determine the supplementary angle's measure. If the pair form a linear pair, label the illustration **linear**.

❶

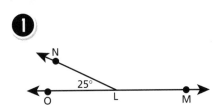

m∠OLN = _____

m∠NLM = _____

_____

❷

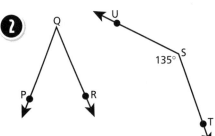

m∠PQR = _____

m∠UST = _____

_____

❸

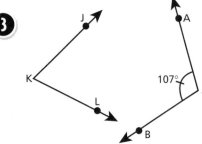

m∠JKL = _____

m∠AJB = _____

_____

❹

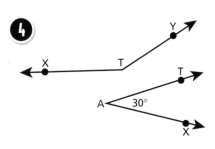

m∠XTY = _____

m∠TAX = _____

_____

❺

m∠TOP = _____

m∠MOT = _____

_____

❻

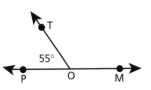
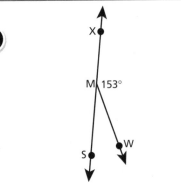

m∠XMW = _____

m∠SMW = _____

_____

Geometry © 2004 Creative Teaching Press

Find Values of Angles

Supplementary or linear pairs have sums of 180°. Complementary pairs have sums of 90°. Use the given information to find the missing angle measures.

1 ∠A and ∠B are supplementary angles.
m∠A = 32°

m∠B= _____

Why?_____

2 ∠C and ∠D are supplementary angles.
m∠C = 102°

m∠D = _____

Why?_____

3 ∠X and ∠Y are linear angles.
m∠X = 98°

m∠Y= _____

Why?_____

4 ∠G and ∠H are complementary angles.
m∠G = 41°

m∠H= _____

Why?_____

5 ∠M and ∠P are supplementary angles.
m∠M = 43°

m∠P = _____

Why?_____

6 ∠J and ∠Y are linear angles.
m∠Y = 93°

m∠J = _____

Why?_____

7 ∠T and ∠V are linear angles.
m∠T = 57°

m∠V = _____

Why?_____

8 ∠F and ∠O are supplementary angles.
m∠F = 111°

m∠O = _____

Why?_____

9 ∠E and ∠R are complementary angles.
m∠E = 16°

m∠R = _____

Why?_____

10 ∠S and ∠U are supplementary angles.
m∠S = 18°

m∠U = _____

Why?_____

Geometry © 2004 Creative Teaching Press

Name _____ Date _____

Vertical Angles

Vertical angles can be thought of as opposite angles. Their sides form two pairs of opposite rays. Vertical angles are the nonadjacent angles formed when two lines intersect.

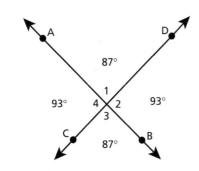

Line AB and line CD intersect.

Angle 1 and angle 3 are vertical angles.
Angle 2 and angle 4 are vertical angles.

Vertical angles are congruent. The angle measure for each vertical angle pair will be the same.
Adjacent angles are supplementary.
∠1 + ∠2 = 180°
∠3 + ∠4 = 180°

Use vertical angles to determine the missing measures. Name the vertical angle pairs.

1

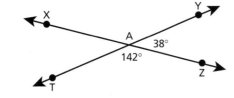

m∠XAY = _____

m∠XAT = _____

Vertical pairs:

_____ and _____

_____ and _____

2

m∠PQN = _____

m∠JQP = _____

Vertical pairs:

_____ and _____

_____ and _____

3

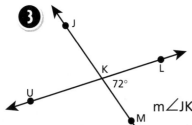

m∠JKL = _____

m∠UKJ = _____

Vertical pairs:

_____ and _____

_____ and _____

4

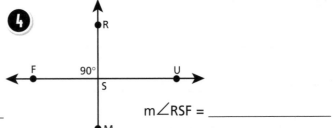

m∠RSF = _____

m∠FSM = _____

Vertical pairs:

_____ and _____

_____ and _____

Geometry © 2004 Creative Teaching Press

Identify Types of Angles

Use the diagrams to name the following angles.

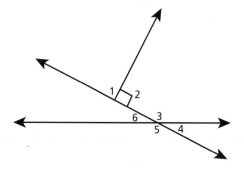

1 Two vertical pairs

_____

2 Two right angles

_____

3 Two acute angles

_____

4 Three linear pairs

_____

5 Two vertical pairs

_____

6 Three adjacent pairs

_____

7 Two obtuse angles

_____

8 Two linear pairs

_____

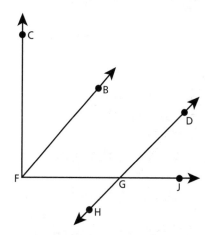

Geometry © 2004 Creative Teaching Press

Find Missing Angle Measurements

Use your understanding of complementary, supplementary, and linear angles to find the missing measures.

∠10 and ∠11 are complementary angles.

1 If m∠10 = 32°, then m∠11 = _____

2 If m∠10 = 63°, then m∠11 = _____

3 If m∠10 = 11°, then m∠11 = _____

∠14 and ∠15 are supplementary angles.

4 If m∠14 = 68°, then m∠15 = _____

5 If m∠14 = 111°, then m∠15 = _____

6 If m∠14 = 87°, then m∠15 = _____

∠M and ∠P are linear angles.

7 If m∠M = 67°, then m∠P = _____

8 If m∠M = 132°, then m∠P = _____

9 If m∠M = 44°, then m∠P = _____

∠5 and ∠6 are complementary angles. ∠6 and ∠7 are supplementary angles. All are nonadjacent.

10 If m∠5 = 34°, then m∠6 = _____, and m∠7 = _____

11 If m∠6 = 50°, then m∠5 = _____, and m∠7 = _____

12 If m∠7 = 132°, then m∠6 = _____, and m∠5 = _____

Geometry © 2004 Creative Teaching Press

Name _____ Date _____

Determine Missing Angles

Use what you know about complementary, supplementary, and vertical angles to determine the missing angle measures.

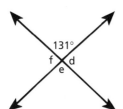

Vertical angles are congruent.
Therefore, e = 131°.

The sum of linear supplemental angles is equal to 180°.
Therefore, $e + f$ = 180°.
$131° + f$ = 180°.
f = 49°

Angles f and d are vertical.
Therefore, d = 49°.

1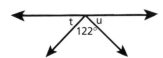

$\angle t \cong \angle u$

$m\angle t =$ _____

$m\angle u =$ _____

2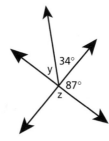

$m\angle y =$ _____

$m\angle z =$ _____

3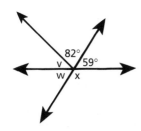

$m\angle v =$ _____

$m\angle x =$ _____

$m\angle w =$ _____

4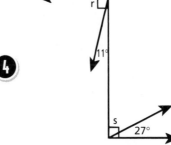

$m\angle r =$ _____

$m\angle s =$ _____

Geometry © 2004 Creative Teaching Press

Name _____ Date _____

Angles and Algebra

Use algebraic equations to find a missing angle measure.

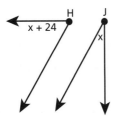

∠J and ∠H are complementary.

m∠J = x
m∠H = x + 24

x + x + 24 = 90°
2x + 24 = 90°
2x = 66°
x = 33°
m∠J = 33°, m∠H = 57°

Write and solve an equation to find the value of *x*.

 1

∠F and ∠G are supplementary.

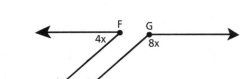

m∠F = _____

m∠G = _____

_____ = 180°

m∠F = _____, m∠G = _____

---

2 ∠L and ∠M are complementary.

m∠L = _____

m∠M = _____

_____ = 90°

m∠L = _____, m∠M = _____

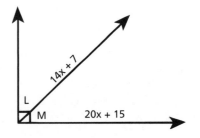

Geometry © 2004 Creative Teaching Press

More Applications of Algebra

Use properties of angles to write and solve algebraic equations.

1

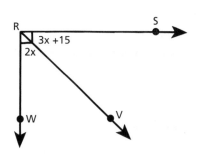

∠SRV and ∠WRV are complementary.

m∠SRV = _____

m∠WRV = _____

_____

m∠SRV = _____, m∠WRV = _____

2

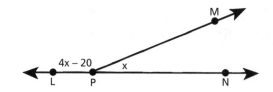

∠MPN and ∠LPM are linear.

m∠MPN = _____

m∠LPM = _____

_____

m∠MPN = _____, m∠LPM = _____

3

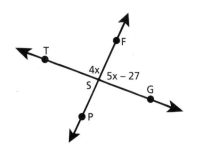

$\overleftrightarrow{TG}$ and $\overleftrightarrow{FP}$ intersect at Point S.

m∠FSG = _____

m∠FST = _____

_____

m∠FSG = _____, m∠FST = _____

m∠PSG = _____, m∠PST = _____

Geometry © 2004 Creative Teaching Press

Vocabulary to Describe Intersections

The vocabulary used to describe lines and intersections helps determine their properties and their relationships to other lines. Match each term in the word box to its definition.

> transversal   intersect   perpendicular   corresponding   alternate
> parallel   bisector   interior   exterior   same side

1 _____ This means to "meet or cross."

2 _____ This describes lines that intersect to form right angles.

3 _____ This describes angles that are on the same side of a transversal and on the same side of given lines.

4 _____ This is a line that intersects two or more other lines at different points.

5 _____ This refers to angles that are on the outside of given lines.

6 _____ This describes something that is always the same distance apart. Lines described as these lie in the same plane and do not intersect.

7 _____ This describes angles on the opposite sides of the transversal on either the outside or inside of given lines.

8 _____ This is a line that divides a geometric form in half.

9 _____ This describes angles that are located on the same side of a transversal.

10 _____ This describes angles that are on the inside of given lines.

Geometry © 2004 Creative Teaching Press

Perpendicular Lines

When two lines intersect, four angles are formed. When two lines intersect to form a right angle, the lines are perpendicular.

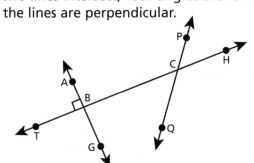

Lines AG, TH, and PQ cross to form angles.
Angle PCB forms an obtuse angle.
Line PQ and line TH are not perpendicular.

Angle ABT forms a 90° or right angle.
Line AG and line TH are perpendicular.
$\overleftrightarrow{AG} \perp \overleftrightarrow{TH}$

Use the diagram to complete.

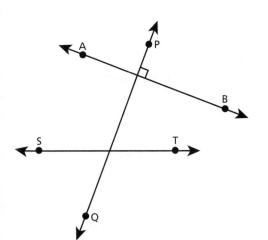

1 Which lines intersect to form angles?

_____

2 Mark the intersection of $\overleftrightarrow{AB}$ and $\overleftrightarrow{PQ}$ as Point E.

3 Mark the intersection of $\overleftrightarrow{ST}$ and $\overleftrightarrow{PQ}$ as Point F.

4 What kind of angle is ∠PEB?

_____

5 What is the relationship of $\overleftrightarrow{PQ}$ and $\overleftrightarrow{AB}$?

_____

6 Mark m∠SFQ = 69°. Using vertical angles, find these measures:

m∠PFT = _____   m∠QFT = _____

m∠SFP = _____

7 What is the relationship of $\overleftrightarrow{PQ}$ and $\overleftrightarrow{ST}$?

_____

Geometry © 2004 Creative Teaching Press

Name _____ Date _____

Parallel Lines

Parallel lines are those lines on the same plane but do not intersect.

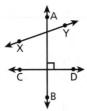

These lines intersect.
$\overleftrightarrow{AB}$ and $\overleftrightarrow{CD}$ are perpendicular.
Since all lines intersect, none
are parallel.

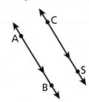

These lines are on the same
plane and do not intersect.
They are parallel. Parallel
lines are marked:

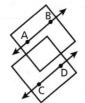

These lines do not intersect,
but they are not on the
same plane. They are not
parallel.

For each diagram, name the lines that are intersecting, parallel, or perpendicular.

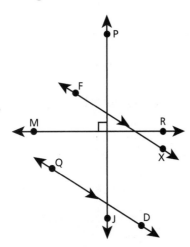

1 Name two lines that intersect.

_____

2 Name two lines that are parallel.

_____

3 Name two lines that are perpendicular.

_____

4 Name two lines that intersect.

_____

5 Name two lines that are parallel.

_____

6

Name two lines that are perpendicular.

_____

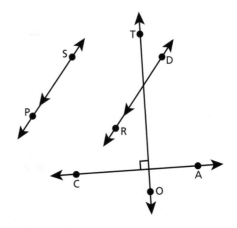

Geometry © 2004 Creative Teaching Press

Name _____ Date _____

Transversals and Corresponding Angles

When a line called a **transversal** intersects two other lines, pairs of angles, called **corresponding angles,** are formed. Corresponding angles are on the same side of a transversal and the same side of a given line.

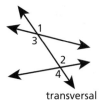

Two nonparallel lines intersected by a transversal.

∠1 corresponds to ∠2
∠3 corresponds to ∠4

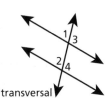

When parallel lines are intersected by a transversal, corresponding angles are congruent.

∠1 ≅ ∠2
∠3 ≅ ∠4

Use the diagrams to name transversals, lines, and corresponding angles.

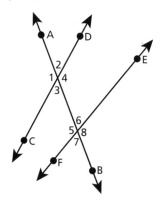

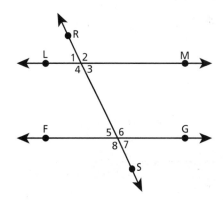

❶ Which line is the transversal?

_____

❷ Which lines are intersected by the transversal?

_____

❸ Name the pairs of corresponding angles.

_____

_____

_____

❹ Which line is the transversal?

_____

❺ Which lines are intersected by the transversal?

_____

❻ Name the pairs of congruent angles.

_____

_____

_____

_____

Geometry © 2004 Creative Teaching Press

Name _____ Date _____

Interior Angles

Interior angles are formed when a transversal intersects two lines.

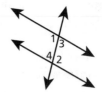

Same-side interior angles are on the same side of the transversal and on the inside of the given lines.

∠1 and ∠2 are same-side interior angles.

∠3 and ∠4 are same-side interior angles.

Alternate interior angles are on opposite sides of the transversal and on the inside of the given lines.

∠1 and ∠2 are alternate interior angles.

∠3 and ∠4 are alternate interior angles.

Use the diagrams to name transversals, lines, and interior angles.

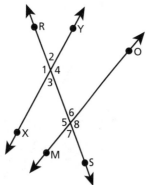

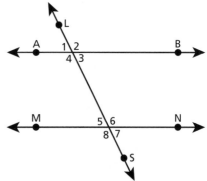

1 Which line is the transversal?

_____

2 Which lines are intersected by the transversal?

_____

3 Name the pairs of same-side interior angles.

_____

4 Name the pairs of alternate interior angles.

_____

5 Which line is the transversal?

_____

6 Which lines are intersected by the transversal?

_____

7 Name the pairs of same-side interior angles.

_____

8 Name the pairs of alternate interior angles.

_____

Geometry © 2004 Creative Teaching Press

Name _____ Date _____

Transversals and Exterior Angles

Exterior angles are formed when a transversal intersects two lines.

Same-side exterior angles are on the same side of the transversal and on the outside of the given lines.

∠1 and ∠2 are same-side exterior angles.

∠3 and ∠4 are same-side exterior angles.

Alternate exterior angles are on opposite sides of the transversal and on the outside of the given lines.

∠1 and ∠2 are alternate exterior angles.

∠3 and ∠4 are alternate exterior angles.

Use the diagrams to name transversals, lines, and exterior angles.

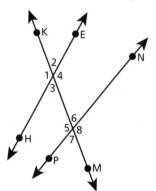

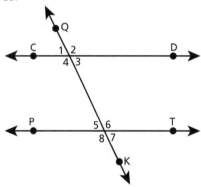

1 Which line is the transversal?

_____

5 Which line is the transversal?

_____

2 Which lines are intersected by the transversal?

_____

6 Which lines are intersected by the transversal?

_____

3 Name the pairs of same-side exterior angles.

_____

7 Name the pairs of same-side exterior angles.

_____

4 Name the pairs of alternate exterior angles.

_____

8 Name the pairs of alternate exterior angles.

_____

Geometry © 2004 Creative Teaching Press

Identify Transversal Angles

For each diagram, identify each type of corresponding angle.

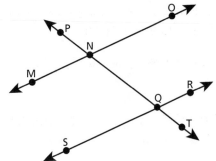

1 Name the alternate exterior angles.

_____

2 Name the alternate interior angles.

_____

3 Name the same-side interior angles.

_____

4 Name the same-side exterior angles.

_____

5 Name two pairs of corresponding angles.

_____

6 Name the alternate exterior angles.

_____

7 Name the alternate interior angles.

_____

8 Name the same-side exterior angles.

_____

Write **true** or **false** to answer questions 9–12.

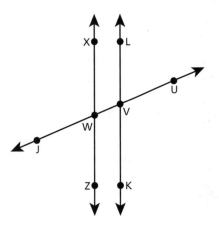

9 ∠XWJ and ∠LVU are same-side exterior angles. _____

10 ∠LVU and ∠JWZ are alternate interior angles. _____

11 ∠XWU and ∠UVK are alternate exterior angles. _____

12 ∠XWU and ∠LVJ are same-side interior angles. _____

Geometry © 2004 Creative Teaching Press

Congruent Transversal Angle Measures

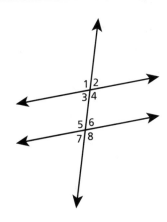

If two parallel lines are crossed by a transversal, then certain angles are congruent.

- Corresponding angles are congruent.
  $\angle 1 \cong \angle 5$

- Alternate interior angles are congruent.
  $\angle 3 \cong \angle 6$

- Alternate exterior angles are congruent.
  $\angle 2 \cong \angle 7$

Two parallel lines intersected by a transversal

Use the rules above to find the missing angle measures.

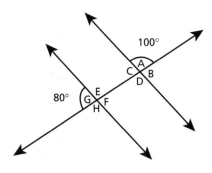

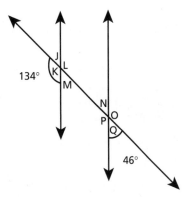

1 $\angle B = $ _____

2 $\angle C = $ _____

3 $\angle D = $ _____

4 $\angle E = $ _____

5 $\angle F = $ _____

6 $\angle H = $ _____

7 $\angle J = $ _____

8 $\angle L = $ _____

9 $\angle M = $ _____

10 $\angle N = $ _____

11 $\angle O = $ _____

12 $\angle P = $ _____

Geometry © 2004 Creative Teaching Press

Name _____ Date _____

Supplementary Transversal Angle Measures

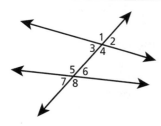

Two lines intersected by a transversal

If two lines are crossed by a transversal, then certain angles are supplementary.

- The pairs of exterior angles are supplementary.
  $\angle 1 + \angle 2 = 180°$

- The pairs of interior angles are supplementary.
  $\angle 5 + \angle 6 = 180°$

- Remember: Vertical angles are congruent.
  $\angle 1 \cong \angle 4$

Use the rules above to find the missing angle measures.

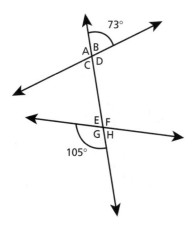

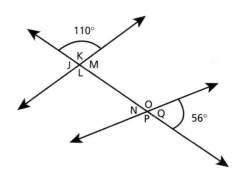

1 ∠A = _____

2 ∠C = _____

3 ∠D = _____

4 ∠E = _____

5 ∠F = _____

6 ∠H = _____

7 ∠J = _____

8 ∠L = _____

9 ∠M = _____

10 ∠N = _____

11 ∠O = _____

12 ∠P = _____

Geometry © 2004 Creative Teaching Press

More Transversal Angles

Use the rules for transversal angles to find the missing angle measures.

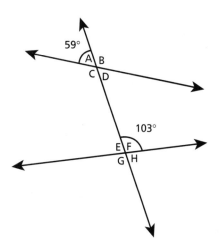

1 ∠B = _____

2 ∠C = _____

3 ∠D = _____

4 ∠E = _____

5 ∠G = _____

6 ∠H = _____

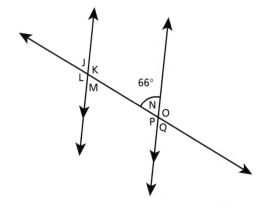

7 ∠J = _____

8 ∠K = _____

9 ∠L = _____

10 ∠M = _____

11 ∠O = _____

12 ∠P = _____

13 ∠Q = _____

Name _____ Date _____

Midpoint Bisectors

A **bisector** divides a geometric figure into two congruent halves. A bisector can be a point, a line, or a plane. The bisector of a line segment is its midpoint.

Point M is the midpoint on this line segment.

A   M   C

1. Find the midpoint coordinate of a line segment by using its number line coordinates and this formula:

$$\frac{a + b}{2}$$

2. The coordinate of Q is 4 cm.
   The coordinate of T is 7 cm.

$$\frac{4 + 7}{2} = \frac{11}{2} = 5.5$$

3. The coordinate of the midpoint is 5.5 cm.

Use the number line to find the midpoints for the following segments.

A   B   C   D   E   F   G   H   I   J   K   L   M   N   O   P   Q   R   S   T   U   V   W   X   Y   Z
-12 -11 -10 -9   -8   -7   -6   -5   -4   -3   -2   -1   0   1   2   3   4   5   6   7   8   9   10   11   12   13

1 $\overline{IS}$

midpoint
coordinate = _____

2 $\overline{JW}$

midpoint
coordinate = _____

3 $\overline{GP}$

midpoint
coordinate = _____

4 $\overline{PV}$

midpoint
coordinate = _____

5 $\overline{BN}$

midpoint
coordinate = _____

6 $\overline{DP}$

midpoint
coordinate = _____

7 $\overline{KU}$

midpoint
coordinate = _____

8 $\overline{JZ}$

midpoint
coordinate = _____

9 $\overline{LX}$

midpoint
coordinate = _____

Geometry © 2004 Creative Teaching Press

Angle Bisectors

An **angle bisector** is a segment or ray in the interior of an angle that divides it into two congruent angles.

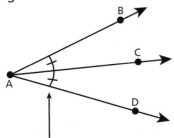

The matching arcs identify congruent angles.

Ray AC bisects angle BAD.
$\overrightarrow{AC}$ bisects ∠BAD

m∠BAC = m∠CAD

If m∠BAD = 42°,
then m∠BAC = 21°
and m∠CAD = 21°.

Use the diagrams to answer the questions about angle bisectors.

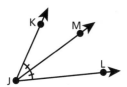

1 _____ bisects ∠KJL.

2 _____ = _____

3 If m∠MJL = 30°, then

m∠KJL = _____ and

m∠KJM = _____

4 _____ bisects ∠MPO.

5 _____ = _____

6 If m∠MPO = 48°, then

m∠MPN = _____ and

m∠NPO = _____

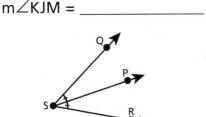

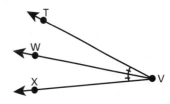

7 _____ bisects ∠QSR.

8 If m∠PSR = 28°, then

m∠QSR = _____ and

m∠QSP = _____

9 _____ = _____

10 If m∠TVW = 16°, then

∠WVX = _____ and

∠TVX = _____

Name _____ Date _____

Algebra and Intersecting Lines

Write an algebraic equation to find the value of *x*.

Solve for *x*.
Find the m∠A and m∠B.

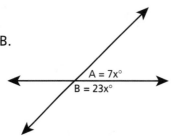

A = 7x°
B = 23x°

Vertical angles are congruent.
Linear pairs = 180°

7x + 23x = 180°
30x = 180°

x = 6°
m∠A: 7x° = 42°   m∠B: 23x = 138°

Use what you know about transversal angles to write an algebraic equation. Solve for *x*. Then find the measure of the given angles.

1

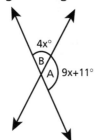

4x°
B
A 9x+11°

_____ + _____ = 180°

x = _____

m∠A: 9x + 11 = _____

m∠B: 4x = _____

2

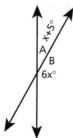

x+5°
A
B
6x°

_____ + _____ = 180°

x = _____

m∠A = _____

m∠B = _____

3

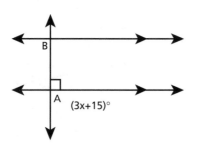

B

A (3x+15)°

x = _____

m∠A = _____

m∠B = _____

4

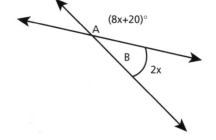

(8x+20)°
A
B
2x

x = _____

m∠A = _____

m∠B = _____

Geometry © 2004 Creative Teaching Press

Algebra and Transversals

Use what you know about transversal angles to write an algebraic equation. Solve for *x*. Then find the measure of the given angles.

1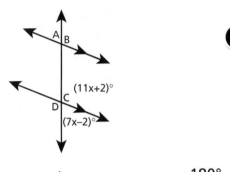

_____ + _____ = 180°

x = _____

m∠A = _____

m∠B = _____

m∠C = _____

m∠D = _____

2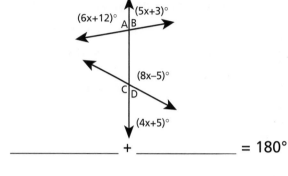

_____ + _____ = 180°

_____ + _____ = 180°

x = _____

m∠A = _____

m∠B = _____

m∠C = _____

m∠D = _____

3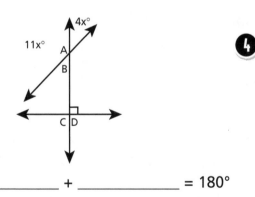

_____ + _____ = 180°

x = _____

m∠A = _____

m∠B = _____

m∠C = _____

m∠D = _____

4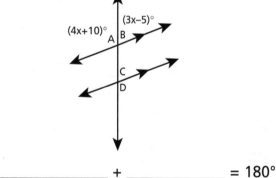

_____ + _____ = 180°

x = _____

m∠A = _____

m∠B = _____

m∠C = _____

m∠D = _____

Geometry © 2004 Creative Teaching Press

Triangles

Triangles have special properties that distinguish them from other polygons. Match each term in the word box to its definition.

triangle	isosceles	scalene	acute	equiangular
obtuse	right	hypotenuse	altitude	legs
base	equilateral			

1 _____ This describes a triangle with interior angles that each measure less than 90°.

2 _____ This is the name for a polygon with three sides.

3 _____ Also known as height, this is shown with a line segment drawn from a vertex and perpendicular to the opposite side or to the line containing the opposite side.

4 _____ This is a triangle with exactly one interior angle that measures more then 90°.

5 _____ This is a triangle that has no congruent sides.

6 _____ Found in a right triangle, this is the side opposite the right angle.

7 _____ This describes a triangle in which all of the angles are the same measure.

8 _____ These describe two congruent sides of an isosceles triangle or two sides of a right triangle that form the right angle.

9 _____ This describes a triangle with one interior angle that measures exactly 90°.

10 _____ This describes the noncongruent side of an isosceles triangle that is not also an equilateral triangle.

11 _____ This is a triangle with two congruent sides.

12 _____ This is a triangle in which all sides are the same length.

Geometry © 2004 Creative Teaching Press

Classify Triangles by Sides

Triangles are classified by their sides into three categories.

scalene triangle

All sides have different lengths.

isosceles triangle

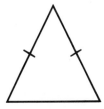

Two sides are the same length. The tick marks show two congruent sides.

equilateral triangle

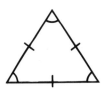

All three sides are the same length. The tick marks show three congruent sides.

Write **scalene, isosceles,** or **equilateral** to classify each triangle.

1

Type: _____

2

Type: _____

3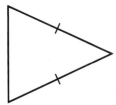

Type: _____

4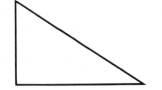

Type: _____

5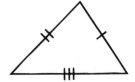

Type: _____

6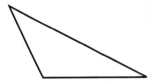

Type: _____

7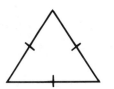

Type: _____

8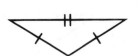

Type: _____

9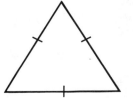

Type: _____

Geometry © 2004 Creative Teaching Press

Classify Triangles by Angles

Triangles are classified by their angles into four categories.

acute triangle	**obtuse triangle**	**right triangle**	**equiangular triangle**

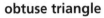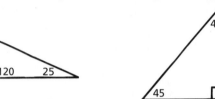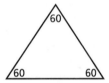

All three interior angles are acute.

Exactly one interior angle is obtuse.

Exactly one interior angle measures 90°.

All interior angles are congruent.

Write **acute, obtuse, right,** or **equiangular** to classify each triangle.

1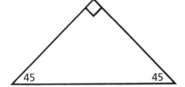

Type: _____

2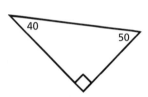

Type: _____

3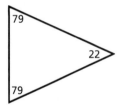

Type: _____

4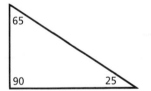

Type: _____

5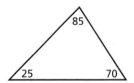

Type: _____

6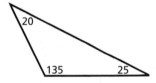

Type: _____

7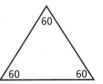

Type: _____

8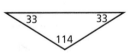

Type: _____

9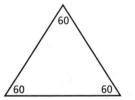

Type: _____

Geometry © 2004 Creative Teaching Press

Name _____ Date _____

What Triangle Is It?

Because triangles can be classified by angles and by sides, each triangle has at least two names. Use the terms in the word box to classify each triangle. Terms will be used more than once.

scalene   isosceles   equilateral   acute   obtuse
right   equiangular

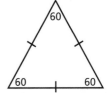

1 _____

2 _____

3 _____

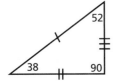

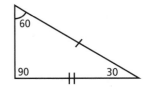

4 _____

5 _____

6 _____

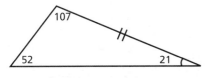

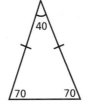

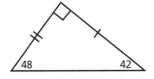

7 _____

8 _____

9 _____

Geometry © 2004 Creative Teaching Press

Find Missing Triangle Measures

Each triangle has three angles. The sum of the three angles is 180°. Isosceles triangles have two congruent angles. Equiangular triangles have three 60° angles. Use this information to determine the missing angle measures.

①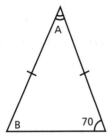

m∠A = _____

m∠B = _____

②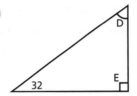

m∠D = _____

m∠E = _____

③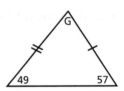

m∠G = _____

④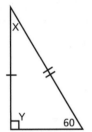

m∠X = _____

m∠Y = _____

⑤

m∠M = _____

⑥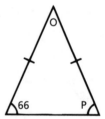

m∠O = _____

m∠P = _____

⑦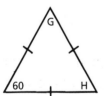

m∠G = _____

m∠H = _____

⑧

m∠R = _____

m∠T = _____

⑨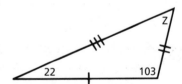

m∠Z = _____

Geometry © 2004 Creative Teaching Press

Name _____ Date _____

Determine Triangle Measures

Use what you know about triangles to write the missing angles.

1 A right triangle has an angle that is 50°. What are the measures of the other two angles?

_____

2 A scalene triangle has an angle that measures 47° and a second angle that measures 88°. What is the measure of the third angle?

_____

3 An equiangular triangle has an angle that measures 60°. What are the measures of the other two angles?

_____

4 An acute triangle has an angle that measures 89° and a second angle measuring 37°. What is the measure of the third angle?

_____

5 A right triangle has an angle that is 60°. What are the measures of the other two angles?

_____

6 An obtuse triangle has one angle measuring 28° and a second angle measuring 42°. What is the measure of the third angle?

_____

7 A scalene triangle has an angle that measures 17° and a second angle that measures 55°. What is the measure of the third angle?

_____

8 An acute triangle has an angle that measures 73° and a second angle measuring 71°. What is the measure of the third angle?

_____

9 A right triangle has an angle that is 35°. What are the measures of the other two angles?

_____

10 A right triangle has an angle that is 58°. What are the measures of the other two angles?

_____

11 An equilateral triangle has an angle that measures 60°. What are the measures of the other two angles?

_____

12 An obtuse triangle has one angle measuring 44° and a second angle measuring 36°. What is the measure of the third angle?

_____

Geometry © 2004 Creative Teaching Press

Segments in Triangles

Triangles contain segments that have special properties.

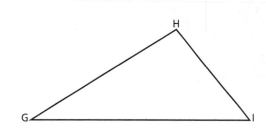

Angle G is opposite segment HI.
∠H is opposite $\overline{GI}$.
∠I is opposite $\overline{GH}$.

Angle G is adjacent to segments GH and GI.
∠H is adjacent to $\overline{HG}$ and $\overline{HI}$.
∠I is adjacent to $\overline{IG}$ and $\overline{IH}$.

Use the diagrams to determine opposite and adjacent geometric features.

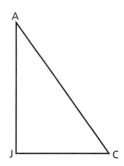

1 ∠A is opposite segment _____.

2 ∠J is opposite segment _____.

3 ∠C is adjacent to _____ and _____.

4 ∠A is adjacent to _____ and _____.

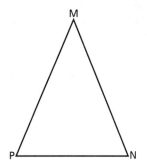

5 ∠M is opposite segment _____.

6 ∠P is opposite segment _____.

7 ∠P is adjacent to _____ and _____.

8 ∠N is adjacent to _____ and _____.

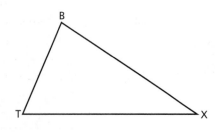

9 _____ is opposite $\overline{BX}$.

10 _____ is opposite $\overline{TX}$.

11 _____ is adjacent to $\overline{BT}$ and $\overline{TX}$.

12 _____ is adjacent to $\overline{XT}$ and $\overline{BX}$.

Geometry © 2004 Creative Teaching Press

Name _____ Date _____

Segments in Isosceles Triangles

The parts of an isosceles triangle have special relationships and names.

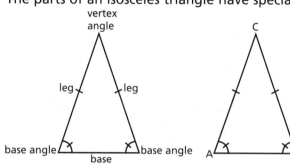

The legs are congruent.
The base angles are congruent.

If base angle A is 50°, then base angle B must also be 50°.

If leg $\overline{AC}$ has a measure of 3 cm, then leg $\overline{BC}$ must also be 3 cm.

Use the diagrams to solve.

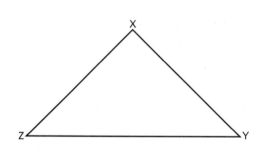

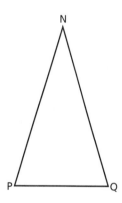

1 Base angles:

_____ ≅ _____

2 Legs:

_____ ≅ _____

3 If ∠Z is 50°, what do you know?

_____

4 If $\overline{XY}$ is 6 cm, what do you know?

_____

5 Base angles:

_____ ≅ _____

6 Legs:

_____ ≅ _____

7 If ∠Q is 76°, what do you know?

_____

8 If $\overline{NQ}$ is 11 cm, what do you know?

_____

Geometry © 2004 Creative Teaching Press

Name _____ Date _____

The Medians of a Triangle

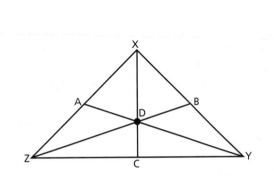

The **median** is a segment from a vertex to the midpoint of the opposite side. Every triangle has three medians.
$\overline{AY}$, $\overline{CX}$, and $\overline{BZ}$ are medians.

The **centroid** is the point at which all three medians are concurrent. It is also the center of gravity for the triangle.
Point D is the centroid.

Each median bisects the opposite side.
$\overline{AZ} \cong \overline{AX}$

Use the diagrams to answer the questions.

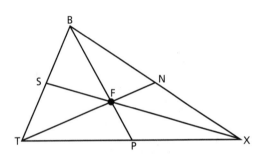

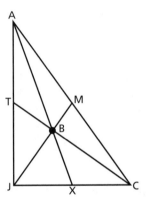

1 Name the medians.

_____ _____ _____

2 What is the centroid of this triangle?

_____

3 Is $\overline{BX}$ a median or a side of this triangle?

_____

4 Is $\overline{PT} \cong \overline{BS}$? _____

5 Is $\overline{PT} \cong \overline{PX}$? _____

6 Name the medians.

_____ _____ _____

7 What is the center of gravity?

_____

8 Is $\overline{TC}$ a median or a side of this triangle?

_____

9 Is $\overline{AT} \cong \overline{TJ}$? _____

10 Is $\overline{MC} \cong \overline{MA}$? _____

Geometry © 2004 Creative Teaching Press

Name _____ Date _____

The Perpendicular Bisectors of a Triangle

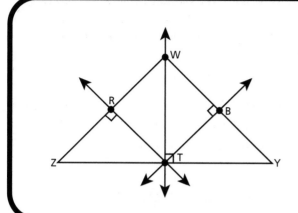

The **perpendicular bisector** is a line, segment, or ray that bisects one side and is perpendicular to it. Every triangle has three perpendicular bisectors. $\overleftrightarrow{RT}$, $\overleftrightarrow{WT}$, and $\overleftrightarrow{BT}$ are perpendicular bisectors.

The common point of all three is the **circumcenter.** It is the same distance from each vertex of the triangle.
Point T is the circumcenter.

Use the diagrams to answer the questions.

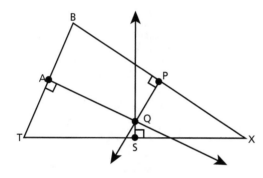

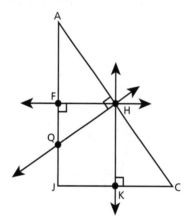

1 Name the perpendicular bisectors.

_____ _____ _____

2 What is the circumcenter of the triangle?

_____

3 Is $\overline{BP}$ part of a perpendicular bisector or a side?

_____

4 If segments are drawn from B to Q and from T to Q, are the segments congruent?

_____

5 Name the perpendicular bisectors.

_____ _____ _____

6 What is the circumcenter of the triangle?

_____

7 Are $\overline{AH}$ and $\overline{HC}$ congruent?

_____

8 If a segment is drawn from J to H, is it congruent to $\overline{CH}$?

_____

Geometry © 2004 Creative Teaching Press

Name _____   Date _____

The Angle Bisectors of a Triangle

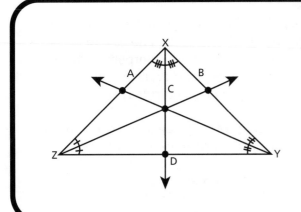

The **angle bisector** is a line, segment, or ray that bisects the angle of a triangle. Every triangle has three angle bisectors. An angle bisector does not necessarily bisect the opposite side.
$\overrightarrow{ZB}$, $\overrightarrow{XD}$, and $\overrightarrow{YA}$ are angle bisectors.

The common point of all three is the **incenter**. The distance is the same from each side to the incenter. Point C is the incenter.

If $\angle XZD = 44°$, then bisector $\angle CZD = 22°$.

Use the diagrams to answer the questions.

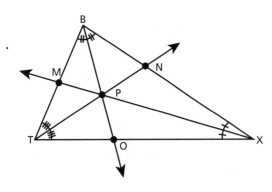

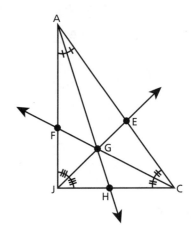

1 Name the angle bisectors.

_____   _____   _____

2 What is the incenter of the triangle?

_____

3 If $\angle BTX = 68°$, name its two bisector angles and their measures.

_____

4 If $\angle BXT = 32°$, name its two bisector angles and their measures.

_____

5 Name the angle bisectors.

_____   _____   _____

6 What is the incenter of the triangle?

_____

7 If $\angle ACJ = 35°$, name its two bisector angles and their measures.

_____

8 If $\angle JCA = 55°$, name its two bisector angles and their measures.

_____

Geometry © 2004 Creative Teaching Press

Name _____ Date _____

The Altitude of a Triangle

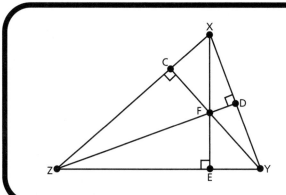

The **altitude** of a triangle is a segment that runs from a vertex perpendicular to the opposite side. The measure of this segment is the height of the triangle. Every triangle has three altitudes. $\overline{ZD}$, $\overline{XE}$, and $\overline{YC}$ are altitudes.

The common point of all three is the **orthocenter**. Point F is the orthocenter.
$\overline{CF} \perp \overline{CZ}$

Use the diagrams to answer the questions.

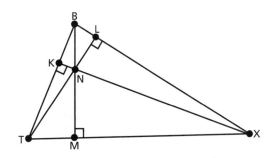

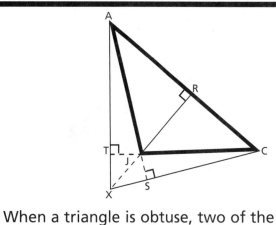

When a triangle is obtuse, two of the altitudes are outside of the triangle. Therefore the orthocenter is outside of the triangle.

1 Name the altitudes.

_____ _____ _____

2 What is the orthocenter of the triangle?

_____

3 What is perpendicular to $\overline{BL}$?

_____

4 What is perpendicular to $\overline{NM}$?

_____

5 Name the altitudes.

_____ _____ _____

6 What is the orthocenter of the triangle?

_____

7 What is perpendicular to $\overline{SC}$?

_____

8 What is perpendicular to $\overline{RC}$?

_____

Geometry © 2004 Creative Teaching Press

Determine Concurrent Lines

Use the terms in the word box to label the type of concurrent lines that have been marked on each triangle.

altitude	median	perpendicular bisector	angle bisector

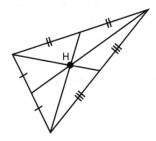

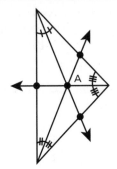

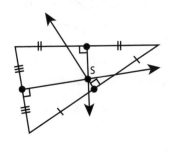

1 _____ **2** _____ **3** _____

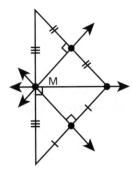

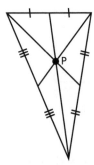

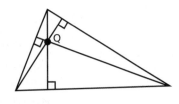

4 _____ **5** _____ **6** _____

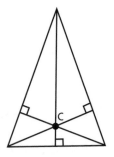

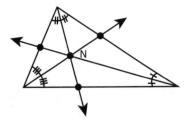

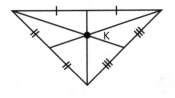

7 _____ **8** _____ **9** _____

Geometry © 2004 Creative Teaching Press

Name _____ Date _____

Quadrilaterals

Quadrilaterals are four-sided polygons. They are classified by the characteristics of their sides or angles. Match each term in the word box to its definition.

rhombus	square	rectangle	quadrilateral	trapezoid
trapezium	kite	parallelogram	base	leg
diagonals	opposite sides	opposite angles		

1 _____ These are line segments that connect two nonadjacent vertices on a quadrilateral.

2 _____ This is a quadrilateral in which both pairs of opposite sides are parallel.

3 _____ This is a parallelogram with four congruent sides, but no right angles.

4 _____ This is a rhombus with four right angles. All four sides and vertices are congruent.

5 _____ This is the general name for a four-sided polygon.

6 _____ These are interior angles of a quadrilateral with no common sides.

7 _____ This is a quadrilateral with exactly one pair of parallel sides.

8 _____ This is a parallelogram with four right angles.

9 _____ These are sides of a quadrilateral that do not have a common endpoint.

10 _____ This is the name used for either of the sides of a trapezoid that are not parallel.

11 _____ This is a quadrilateral with no parallel sides.

12 _____ This is a quadrilateral with no parallel sides, but it has exactly two pairs of adjacent congruent sides. Opposite sides of this polygon are not congruent.

13 _____ This is the name used for either of the parallel sides of a trapezoid.

Geometry © 2004 Creative Teaching Press

Name Quadrilaterals

When classifying quadrilaterals, be as specific as possible. Some quadrilaterals may be classified more than one way. Use the terms in the word box to label each figure with all the classifications that apply. Some terms are used more than once.

| rhombus | square | rectangle | trapezium |
| kite | parallelogram | trapezoid | quadrilateral |

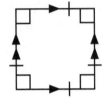

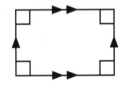

❶ _____ **❷** _____ **❸** _____

_____ _____ _____

_____ _____

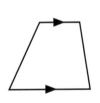

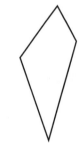

❹ _____ **❺** _____ **❻** _____

_____ _____

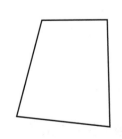

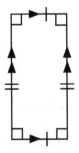

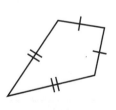

❼ _____ **❽** _____ **❾** _____

_____ _____

_____ _____

Geometry © 2004 Creative Teaching Press

Name _____ Date _____

Use Markings to Show Quadrilaterals

Markings used on geometric figures help show the properties of each figure.

Tick marks are used to show congruence.

Single lines show one congruent set, double lines show a second congruent set, and so on.

Arrowheads indicate that lines are parallel.

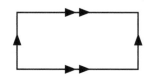

Single arrowheads show one parallel set, double arrowheads show a second parallel set, and so on.

A small square at the vertex of an intersection indicates perpendicular lines.

For each diagram, use the markings to describe its properties. Then classify the quadrilateral based on the properties.

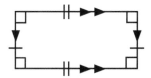

1 _____
_____
_____

This is a _____.

2 _____
_____
_____

This is a _____.

3 _____
_____
_____

This is a _____.

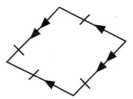

4 _____
_____
_____

This is a _____.

Geometry © 2004 Creative Teaching Press

Name _____   Date _____

Diagonals on Quadrilaterals

A **diagonal** is a line segment that connects two nonadjacent vertices.

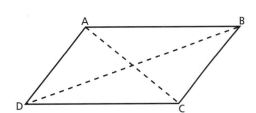

The diagonals divide this parallelogram into four triangles.

This parallelogram has four sides.
To form all the diagonals, draw a line segment to all other vertices. Two of the vertices will already be sides.

Diagonals = $\overline{AC}$ and $\overline{BD}$

Opposite sides = $\overline{AD}$ and $\overline{BC}$

$\overline{AB}$ and $\overline{CD}$

Opposite angles = ∠DAB and ∠DCB

∠ABC and ∠ADC

For each quadrilateral, use a ruler to draw the diagonals. Then answer the questions.

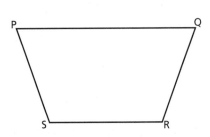

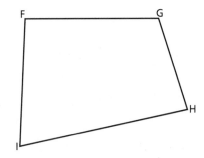

1 Which segments form the diagonals?

_____

4 Which segments form the diagonals?

_____

2 Which segments are opposite sides?

_____

5 Which segments are opposite sides?

_____

3 Which angles are opposite angles?

_____

6 Which angles are opposite angles?

_____

Geometry © 2004 Creative Teaching Press

Parallelograms

A **parallelogram** is a quadrilateral in which both pairs of opposite sides are parallel. There are properties that can help determine if a four-sided figure is a parallelogram.

Opposite sides are congruent.

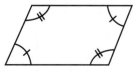

Opposite angles are congruent.

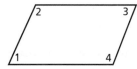

Consecutive angles are supplementary.

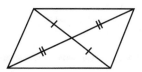

Diagonals bisect each other.

For each parallelogram, name the features that prove each rule.

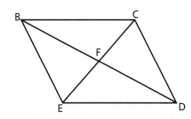

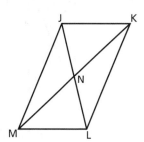

1 Opposite sides are congruent.

_____

2 Opposite angles are congruent.

_____

3 Consecutive angles are supplementary.

_____

_____

4 Diagonals bisect each other.

_____

5 Opposite sides are congruent.

_____

6 Opposite angles are congruent.

_____

7 Consecutive angles are supplementary.

_____

_____

8 Diagonals bisect each other.

_____

Geometry © 2004 Creative Teaching Press

Rectangles

A rectangle has four right angles. According to its properties, a rectangle is also a parallelogram and a quadrilateral.

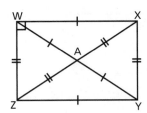

If a parallelogram has one right angle, then it is a rectangle.

Just like a parallelogram, a rectangle has
- opposite sides that are congruent.
- opposite angles that are congruent.
- consecutive angles that are supplementary.
- diagonals that bisect each other.

A rectangle has right angles at its vertices. ∠W is a right angle.

∠W ≅ ∠X ≅ ∠Y ≅ ∠Z

For each diagram, name the features that prove it is a rectangle.

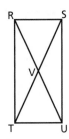
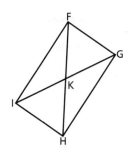

1 Opposite sides are congruent.

_____

2 Opposite angles are congruent.

_____

3 Diagonals bisect each other.

_____

_____

4 Each vertex is a right angle.

_____

_____

5 Opposite sides are congruent.

_____

6 Consecutive angles are supplementary.

_____

7 Diagonals bisect each other.

_____

_____

8 Each vertex is a right angle.

_____

_____

Geometry © 2004 Creative Teaching Press

Name _____ Date _____

Rhombi

Rhombi, plural for **rhombus,** are parallelograms with four congruent sides. A parallelogram is a rhombus if:

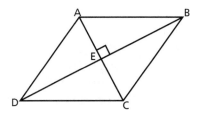

The diagonals are perpendicular to each other.__
$\overline{AC} \perp \overline{BD}$

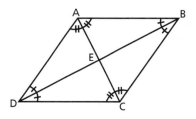

Each diagonal bisects a pair of opposite angles.
∠EDC ≅ ∠ABE ≅ ∠EBC ≅ ∠ADE
∠DCE ≅ ∠ECB ≅ ∠DAE ≅ ∠EAB

A rhombus meets the properties of a parallelogram plus those listed above. For each rhombus, name the features that prove each rule.

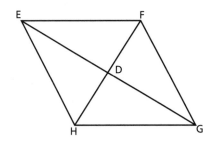

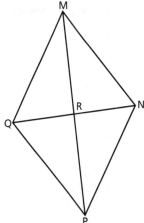

1 Diagonals are perpendicular.

_____

2 Diagonals bisect pairs of opposite angles.

_____

3 All four sides are congruent.

_____

4 Opposite angles are congruent.

_____

5 Diagonals are perpendicular.

_____

6 Diagonals bisect pairs of opposite angles.

_____

7 All four sides are congruent.

_____

8 Opposite angles are congruent.

_____

Geometry © 2004 Creative Teaching Press

Squares

A square can be considered a rhombus with four right angles. It can also be considered a rectangle with four congruent sides. A square is also a quadrilateral and a parallelogram.

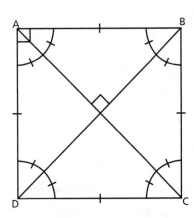

Squares have all the properties of parallelograms, rhombi and rectangles:
- Opposite sides are congruent.
- Opposite angles are congruent.
- Consecutive angles are supplementary.
- Diagonals bisect a pair of opposite angles and each other.

A square also has four congruent sides and four right angles.
- $\angle A \cong \angle B \cong \angle C \cong \angle D$
- $\overline{AB} \cong \overline{BC} \cong \overline{CD} \cong \overline{DA}$

For each figure, name the features that prove it is a square.

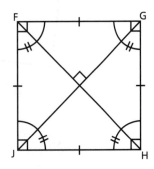

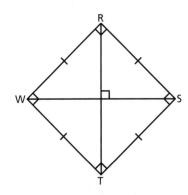

1 _____ is a right angle.

2 All angles are congruent.

_____

3 All four sides are congruent.

_____

4 Diagonals are perpendicular.

_____

5 Diagonals are perpendicular.

_____

6 Diagonals bisect pairs of opposite angles.

_____

7 All four sides are congruent.

_____

8 All four angles are congruent.

_____

Geometry © 2004 Creative Teaching Press

Name _____ Date _____

Trapezoids

A **trapezoid** is a quadrilateral with exactly one pair of parallel sides. There are special names for the sides and angles of a trapezoid.

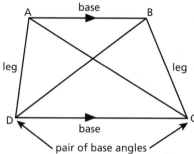

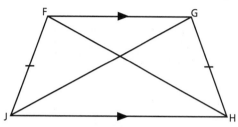

The parallel sides of a trapezoid are the bases. The legs are the nonparallel sides. A pair of base angles is formed at the ends of each base.

If the legs are congruent, then the trapezoid is called an isosceles trapezoid.

If the diagonals are congruent, then the figure is an isosceles trapezoid.

For each figure, name the features that prove it is a trapezoid.

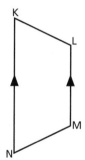

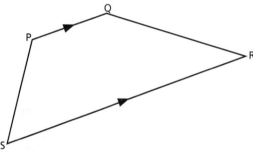

1 Name the bases.

_____

2 Name the legs.

_____

3 Name both pairs of base angles.

_____ _____

4 Is this an isosceles trapezoid? Explain.

_____

_____

5 Name the bases.

_____

6 Name the legs.

_____

7 Name both pairs of base angles.

_____ _____

8 Is this an isosceles trapezoid? Explain.

_____

_____

Geometry © 2004 Creative Teaching Press

Name _____ Date _____

Kites

A **kite** is a quadrilateral with exactly two pairs of adjacent congruent sides.

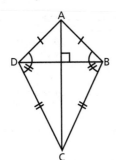

A kite is a trapezium because none of its sides are parallel and it has four sides.

A kite has exactly one pair of opposite congruent angles, but the opposite sides are not congruent.

A kite has perpendicular diagonals.

Write **trapezium** or **kite** to identify each figure. Explain the features that determined the chosen quadrilateral.

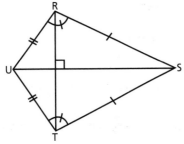

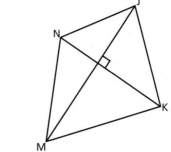

1 It is a _____.

_____

_____

2 It is a _____.

_____

_____

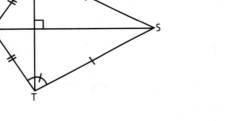

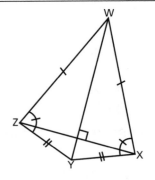

3 It is a _____.

_____

_____

4 It is a _____.

_____

_____

Geometry © 2004 Creative Teaching Press

Define Polygons

Polygons are closed plane figures that can be categorized by their properties. Match each term in the word box to its definition.

concave	convex	regular	interior angle	exterior angle
pentagon	hexagon	heptagon	octagon	nonagon
decagon	dodecagon	irregular		

1 _____ This is a polygon with nine sides.

2 _____ This is a polygon with seven sides.

3 _____ If a line that contains a side of a polygon also contains a point inside the polygon, the polygon is this.

4 _____ This is on the inside of a polygon, formed by its sides.

5 _____ This describes a polygon with eight sides.

6 _____ This is formed by a side of a polygon and an extension of an adjacent side.

7 _____ If none of the lines that contain a side of a polygon also contain a point inside the polygon, the polygon is this.

8 _____ This describes a polygon in which all sides are congruent and all angles are congruent.

9 _____ This describes a polygon with six sides.

10 _____ This is the name for a five-sided polygon.

11 _____ This is a polygon in which at least one pair of sides or angles is not congruent.

12 _____ This is a polygon with ten sides.

13 _____ This is the name for a polygon with twelve sides.

Geometry © 2004 Creative Teaching Press

Name _____ Date _____

Identify Polygons

Polygons are closed plane figures.

A polygon is a closed plane figure formed by line segments that meet only at their endpoints.

The segments are the sides of the polygons. Each side intersects exactly two other sides forming vertices.

Some figures are not polygons.

Open figures are not polygons.
If two sides meet at a point other than an endpoint, it is not a polygon.
All sides must be line segments, not curves.

Write **polygon** or **not a polygon** to classify each figure. If it is not a polygon, explain why.

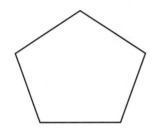 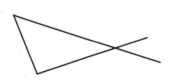 

❶ _____   ❷ _____   ❸ _____

  _____   _____   _____

  _____   _____   _____

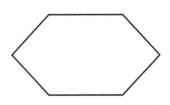 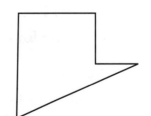

❹ _____   ❺ _____   ❻ _____

  _____   _____   _____

  _____   _____   _____

Geometry © 2004 Creative Teaching Press

Name _____ Date _____

Convex and Concave Polygons

Polygons can be convex or concave.

A **concave** polygon has at least one line containing a side that also contains a point inside the polygon.

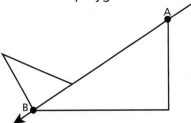

Line AB shows where the points of the line are inside.

A **convex** polygon has no lines containing the sides that contain points inside the polygon.

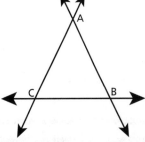

The lines of each side show how none of the points are inside.

Write **convex** or **concave** to classify each polygon. Then draw lines on each to prove your choice.

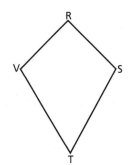

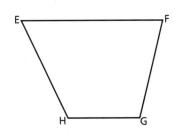

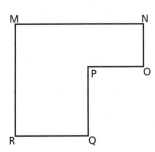

 _____   _____   _____

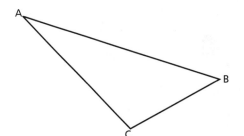

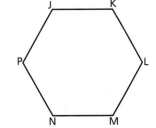

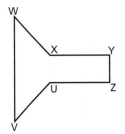

 _____   _____   _____

Geometry © 2004 Creative Teaching Press

Name _____ Date _____

Identify by Sides

Polygons are identified by the number of sides they have.

These are examples of quadrilaterals because they all have four sides.

These are examples of pentagons because they all have five sides.

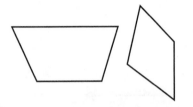 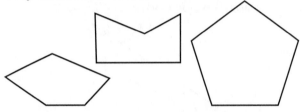

triangle	quadrilateral	pentagon	hexagon	heptagon
nonagon	decagon	dodecagon	octagon	

Write the name of the polygon that reflects the number of sides in each diagram.

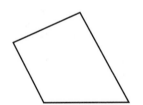 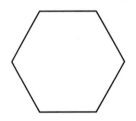 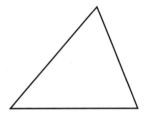

1 _____ **2** _____ **3** _____

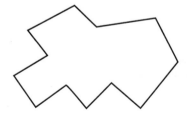 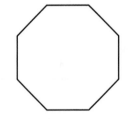 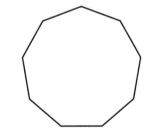

4 _____ **5** _____ **6** _____

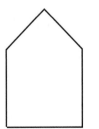 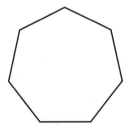 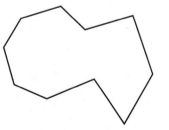

7 _____ **8** _____ **9** _____

Geometry © 2004 Creative Teaching Press

Name _____ Date _____

Diagonals of Polygons

A **diagonal** of a polygon is a line segment that connects two nonadjacent vertices. In any polygon, each vertex can be connected to n − 1 other vertices. Two of these will be sides of the polygon.

A parallelogram has four sides.
If n = the number of sides, then
n − 1 = the number of diagonals.

A pentagon has five sides.
If n = the number of sides, then
n − 1 = the number of diagonals.

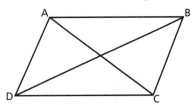

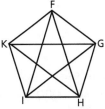

In this parallelogram, each vertex can be connected to 3 other vertices. Two are sides.

In this pentagon, each vertex can be connected to 4 other vertices. Two are sides.

Determine the number of diagonals from each vertex. Then draw the diagonals.

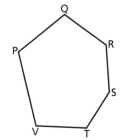

n = _____

n − 1 = _____

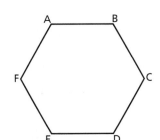

n = _____

n − 1 = _____

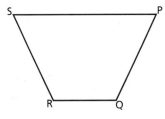

n = _____

n − 1 = _____

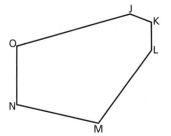

n = _____

n − 1 = _____

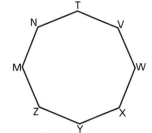

n = _____

n − 1 = _____

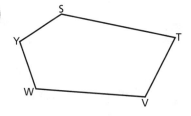

n = _____

n − 1 = _____

Geometry © 2004 Creative Teaching Press

Regular and Irregular Polygons

Polygons can be classified as regular or irregular.

If all the sides and all the angles are congruent, then a polygon is regular.

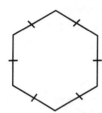

This polygon has 6 sides.
All sides and angles are congruent.
This is a regular hexagon.

If some of the sides and some of the angles are not congruent, then the polygon is irregular.

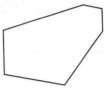

This polygon has 6 sides.
Not all sides and angles are congruent.
This is an irregular hexagon.

Determine the number of sides. Identify the polygon as **regular** or **irregular** and write its name. Name the congruent sides.

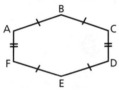

1 Number of sides: _____

2 Name: _____

3 Congruent sides: _____

_____

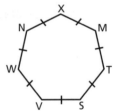

4 Number of sides: _____

5 Name: _____

6 Congruent sides: _____

_____

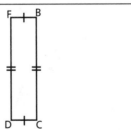

7 Number of sides: _____

8 Name: _____

9 Congruent sides: _____

_____

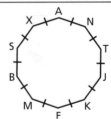

10 Number of sides: _____

11 Name: _____

12 Congruent sides: _____

_____

Geometry © 2004 Creative Teaching Press

Name _____ Date _____

The Sum of All Interior Angles

The angles inside a polygon are called **interior angles.**

To find the sum of the measures of the interior angles:
$180° (n - 2)$ = sum of interior angles
n = the number of sides in the polygon.

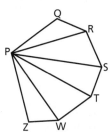

This heptagon has 7 sides. The diagonals divide the figure into 5 different triangles, each totaling 180°.

$180° (7 - 2)$ = sum of interior angles
$180° (5) = 900°$

The sum of the interior angles in this heptagon is 900°.

For each polygon, mark one set of diagonals. Then use the formula to find the sum of the interior angles.

$180° (n - 2)$ = sum of interior angles

_____

_____

sum of interior angles = _____

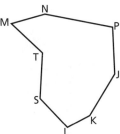

$180° (n - 2)$ = sum of interior angles

_____

_____

sum of interior angles = _____

❸

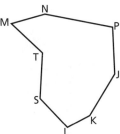

$180° (n - 2)$ = sum of interior angles

_____

_____

sum of interior angles = _____

❹

$180° (n - 2)$ = sum of interior angles

_____

_____

sum of interior angles = _____

Geometry © 2004 Creative Teaching Press

Name _____ Date _____

The Sum of All Exterior Angles

Exterior angles are outside a polygon. An exterior angle is formed by a side of a polygon and an extension of an adjacent side.

The sum of the exterior angles for a polygon is 360°.

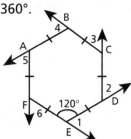

The interior and exterior angles at each vertex form a linear pair.
If the interior angle is known, the exterior angle must be its supplement.

∠FED + ∠1 = 180°
If ∠FED = 120°, then ∠1 = 60°.

60° × 6 exterior angles = 360°

For each polygon, find the linear pair of the marked angles. Prove that the sum of the angles equals 360°.

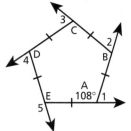

m∠ _____ + m∠ _____ = 180°

m∠ _____ = _____

m∠ _____ = _____

_____ × _____ angles = 360°

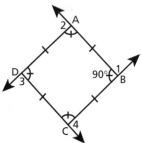

m∠ _____ + m∠ _____ = 180°

m∠ _____ = _____

m∠ _____ = _____

_____ × _____ angles = 360°

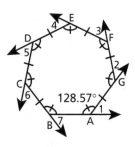

_____

_____

_____

_____ × _____ angles ≈ 360°

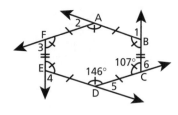

_____

_____

_____

_____

Geometry © 2004 Creative Teaching Press

Name _____ Date _____

Polygon Congruence

Two polygons are **congruent** if their corresponding sides and their corresponding angles are congruent. Congruent polygons are the same shape and size.

These triangles are congruent.

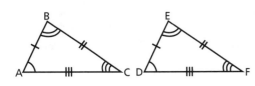

$\triangle ABC \cong \triangle DEF$

If two polygons are congruent, corresponding angles are congruent:

$\angle A \cong \angle D$ $\angle B \cong \angle E$ $\angle C \cong \angle F$

The corresponding sides are also congruent:

$\overline{AB} \cong \overline{DE}$ $\overline{BC} \cong \overline{EF}$ $\overline{AC} \cong \overline{DF}$

Give the corresponding angles and corresponding lines that prove congruency.

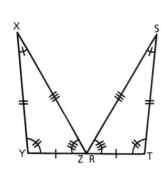

1 $\triangle YXZ \cong \triangle RST$

Corresponding angles: _____

_____

Corresponding sides: _____

_____

2 parallelogram ABCD $\cong$ parallelogram XWYZ

Corresponding angles: _____

_____

Corresponding sides: _____

_____

_____

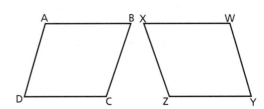

Geometry © 2004 Creative Teaching Press

More Congruent Polygons

Name the corresponding sides and corresponding angles that prove each polygon is congruent.

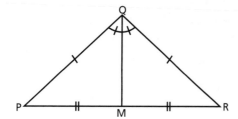

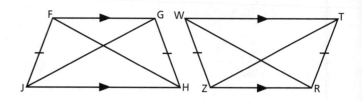

1 ΔQPM ≅ ΔQRM

Corresponding angles: _____

_____

Corresponding sides: _____

_____

2 trapezoid FGHJ ≅ trapezoid WTRZ

Corresponding angles: _____

_____

Corresponding sides: _____

_____

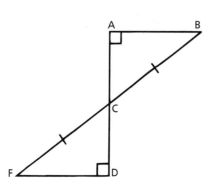

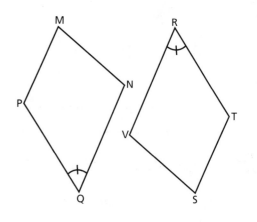

3 ΔABC ≅ ΔCDF

Corresponding angles: _____

_____

Corresponding sides: _____

_____

4 polygon MNQP ≅ polygon RTSV

Corresponding angles: _____

_____

Corresponding sides: _____

_____

Geometry © 2004 Creative Teaching Press

Name _____ Date _____

Find Measures with Congruent Polygons

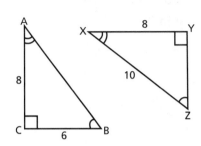

△ABC ≅ △XYZ

Match corresponding sides or corresponding angles. Use known measures to find missing measures.

m$\overline{AB}$ = m$\overline{XZ}$ = 10
m$\overline{BC}$ = m$\overline{YZ}$ = 6
m$\overline{AC}$ = m$\overline{XY}$ = 8

1 △MNO ≅ △RST

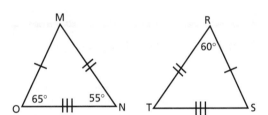

m∠ _____ = m∠ _____ = _____

m∠ _____ = m∠ _____ = _____

m∠ _____ = m∠ _____ = _____

m∠M = _____   m∠T = _____

m∠S = _____

2 polygon JKPL ≅ polygon YMPS

m∠ _____ = m∠ _____ = _____

m∠ _____ = m∠ _____ = _____

m∠ _____ = m∠ _____ = _____

m∠ _____ = m∠ _____ = _____

m∠J = _____   m∠K = _____

m∠Y = _____   m∠M = _____

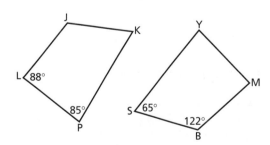

Geometry © 2004 Creative Teaching Press

Name _____ Date _____

Determine Congruence by Sides or Angles

Two ways to determine congruence are the Side-Side-Side or Side-Angle-Side method.

Side-Side-Side (SSS)

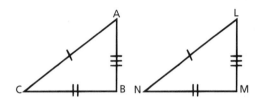

If three sides of one triangle are congruent to three sides of another triangle, then the triangles are congruent.

Side $\overline{AB} \cong \overline{LM}$ Side $\overline{BC} \cong \overline{MN}$ Side $\overline{AC} \cong \overline{LN}$

Side-Angle-Side (SAS)

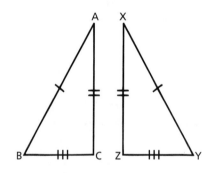

If two sides and the included angle of one triangle are congruent to two sides and the included angle of a second triangle, the triangles are congruent.

Side $\overline{DE} \cong \overline{QR}$ $\angle E \cong \angle R$ Side $\overline{EF} \cong \overline{RS}$

Name the congruent parts for each diagram. Write **SSS** or **SAS** to show how congruence can be determined. Then name the congruent polygons.

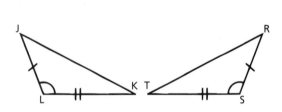

1 Congruent parts:

_____

_____

_____

2 SSS or SAS? _____

3 Δ _____ ≅ Δ _____

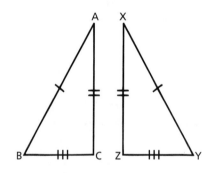

4 Congruent parts:

_____

_____

_____

5 SSS or SAS? _____

6 Δ _____ ≅ Δ _____

Geometry © 2004 Creative Teaching Press

Name _____ Date _____

Use SSS or SAS

Write if congruence can be proved using **SSS** or **SAS** for each diagram. Then write the names of the congruent features.

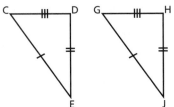

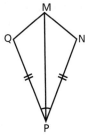

1 _____

2 _____

_____

_____

3 _____

4 _____

_____

_____

5 _____

6 _____

_____

_____

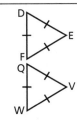

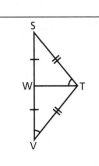

7 _____

8 _____

_____

_____

9 _____

10 _____

_____

_____

11 _____

12 _____

_____

_____

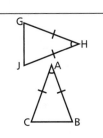

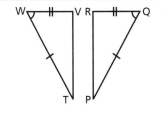

13 _____

14 _____

_____

_____

15 _____

16 _____

_____

_____

17 _____

18 _____

_____

_____

Geometry © 2004 Creative Teaching Press

Similarity

Polygons are **similar** if they are the same shape, but not necessarily the same size. The symbol to show that two polygons are similar is ~.

 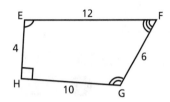

To check proportions, set up corresponding segments as a ratio:

$$\frac{AB}{EF} = \frac{6}{12} \qquad \frac{BC}{FG} = \frac{3}{6} \qquad \frac{CD}{GH} = \frac{5}{10} \qquad \frac{AD}{EH} = \frac{2}{4}$$

If these ratios are equivalent, then the polygons are proportional.

Same shape means the corresponding angles are congruent. The lengths of the corresponding sides are proportional.

$$\frac{6}{12} = \frac{1}{2} \qquad \frac{3}{6} = \frac{1}{2} \qquad \frac{5}{10} = \frac{1}{2} \qquad \frac{2}{4} = \frac{1}{2}$$

Polygon ABCD ~ polygon EFGH

For each pair of polygons, determine the ratios. Prove whether or not they are similar. Name the similar polygons.

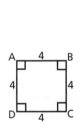 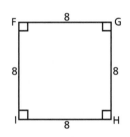 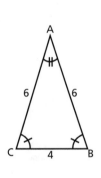

1 Ratios of sides:

4 Ratios of sides:

 2 Are the polygons similar?

 5 Are the polygons similar?

_____

_____

 3 _____ ~ _____

 6 _____

Geometry © 2004 Creative Teaching Press

Check for Similar Polygons

Use corresponding angles and proportions to check for similarity in these polygons.

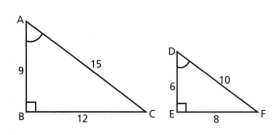

1 Ratios of sides:

2 Are the polygons similar? _____

3 _____ ~ _____

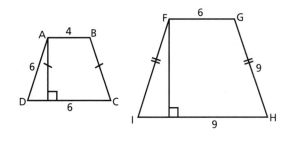

4 Ratios of sides:

5 Are the polygons similar? _____

6 _____

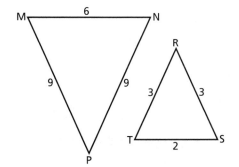

7 Ratios of sides:

8 Are the polygons similar? _____

9 _____

Similar or Congruent?

For each pair of polygons, write **similar, congruent,** or **neither.** Then explain your answer.

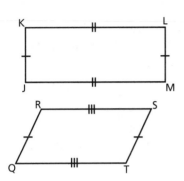

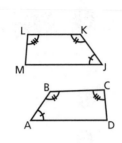

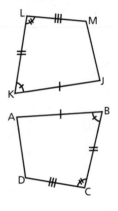

1 _____

_____

2 _____

_____

3 _____

_____

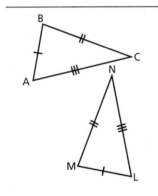

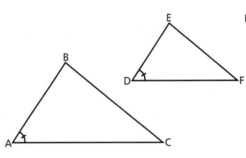

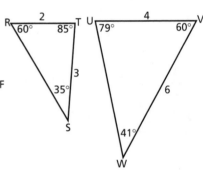

4 _____

_____

5 _____

_____

6 _____

_____

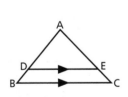

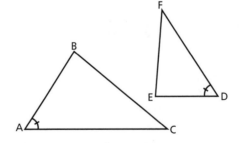

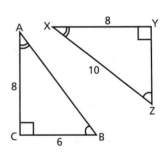

7 _____

_____

8 _____

_____

9 _____

_____

Geometry © 2004 Creative Teaching Press

Name _____ Date _____

Perimeter of Polygons

The **perimeter** of a polygon is the sum of all sides.

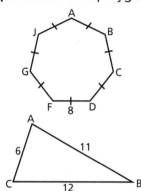

For a regular polygon, the number of sides can be multiplied by the length of a side to find the perimeter.

perimeter = number of sides × side length
8 × 7 = 56 units

For irregular polygons, add all sides to find the perimeter.

perimeter = side AB + side BC + side AC
11 + 12 + 6 = 29 units

Write an equation to find the perimeter of each polygon.

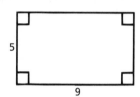

1 _____

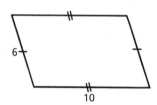

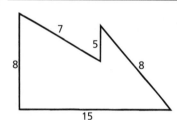

2 _____

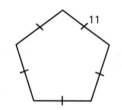

3 _____

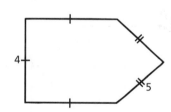

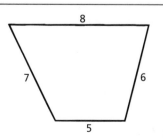

4 _____

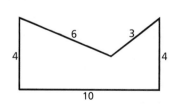

5 _____

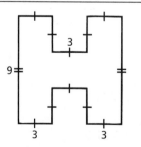

6 _____

7 _____

8 _____

9 _____

Geometry © 2004 Creative Teaching Press

Perimeter with Mixed Measures

Linear measurements may contain two or three different units, such as yards and feet or feet and inches. When computing perimeters with units like these, be sure to add like units and simplify your answers.

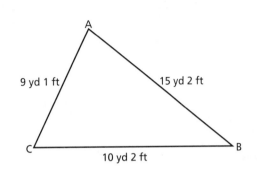

Add like units:
    15 yards 2 feet
    10 yards 2 feet
  + 9 yards 1 foot
    34 yards 5 feet

Simplify the answer: 5 feet > 1 yard
    34 yd 5 ft = 34 yd + 3 ft + 2 ft
               34 yd + 1 yd + 2 ft
               35 yd + 2 ft

The perimeter is 35 yards 2 feet.

Find the perimeters. Show your work. Simplify the answers when needed.

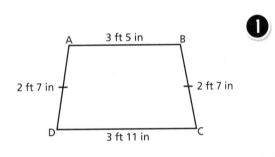

1

The perimeter is _____.

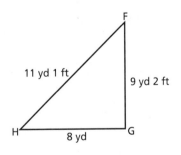

2

The perimeter is _____.

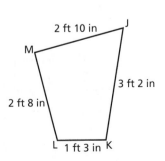

3

The perimeter is _____.

Geometry © 2004 Creative Teaching Press

Name _____ Date _____

Area of Squares and Rectangles

The **area** of a plane figure is the number of square units that a figure covers. Area is measured in square units.

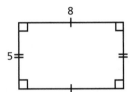

The area of a rectangle is the product of its length and width.

A = lw
A = 5(8)
A = 40 square units

The area of a square is the length of one side squared.

A = s²   where s equals *side*
A = 6²
A = 36 square units

Use the formulas to find the area of these quadrilaterals.

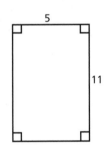

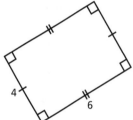

1 Formula:_____

Equation: _____

Area:_____

2 Formula:_____

Equation: _____

Area:_____

3 Formula:_____

Equation: _____

Area:_____

4 Formula:_____

Equation: _____

Area:_____

Geometry © 2004 Creative Teaching Press

Area of Parallelograms and Rhombi

The area of parallelograms and rhombi are found with a formula similar to that for rectangles. Instead of length and width, use base and height.

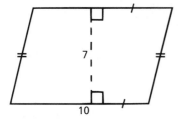

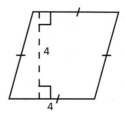

The area of a parallelogram is the product of a base and its corresponding height.

A = bh
A = 10(7)
A = 70 square units

The area of a rhombus is the product of a base and its corresponding height.

A = bh
A = 4(4)
A = 16 square units

Use the formulas to find the area of these quadrilaterals.

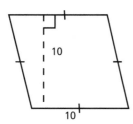

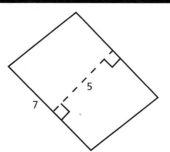

❶ Formula:_____

   Equation: _____

   Area:_____

❷ Formula:_____

   Equation: _____

   Area:_____

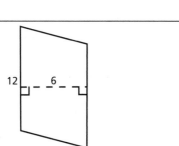

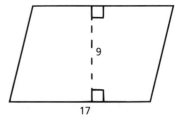

❸ Formula:_____

   Equation: _____

   Area:_____

❹ Formula:_____

   Equation: _____

   Area:_____

Geometry © 2004 Creative Teaching Press

Name _____   Date _____

Area of Trapezoids

The area of a trapezoid is half the product of the height and the sum of the bases.

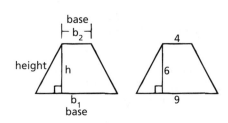

$$A = \frac{1}{2}h(b_1 + b_2)$$

To find the area of a trapezoid, first add the bases. Then multiply that sum by the height. Multiply that product by half.

$$A = \frac{1}{2}(4 + 9)6$$

$$A = \frac{1}{2}(13)6$$

$$A = \frac{1}{2}(78)$$

$$A = 39 \text{ square units}$$

Use the formula to find the area of these trapezoids.

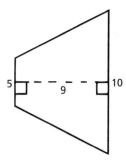

1 A = _____

A = _____

A = _____

Area of trapezoid: _____

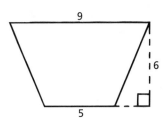

2 A = _____

A = _____

A = _____

Area of trapezoid: _____

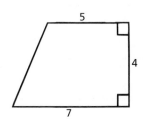

3 A = _____

A = _____

A = _____

Area of trapezoid: _____

Geometry © 2004 Creative Teaching Press

Name _____ Date _____

Area of Triangles

The area of a triangle is half the product of a base and its corresponding height.

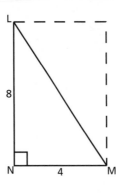

$A = \frac{1}{2}bh$

A congruent triangle can be drawn next to triangle LMN to form a rectangle.
The area of a rectangle is a product of its length and width.
Triangle LMN is only half of this rectangle.
The area of a triangle is the product of the base, the height, and halved.

$A = \frac{1}{2}(8)(4)$

A = 16 square units

Use the formula to find the area of these triangles.

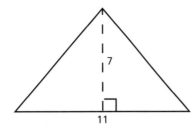

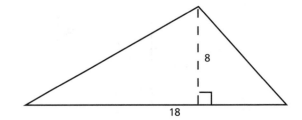

1 Formula:_____

Equation: _____

Area:_____

2 Formula:_____

Equation: _____

Area:_____

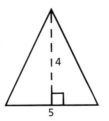

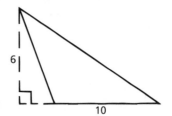

3 Formula:_____

Equation: _____

Area:_____

4 Formula:_____

Equation: _____

Area:_____

Geometry © 2004 Creative Teaching Press

Area of Perpendicular Polygons

Some polygons consist of many perpendicular segments. The area for these polygons can be found by dividing it up by known sides into smaller rectangles.

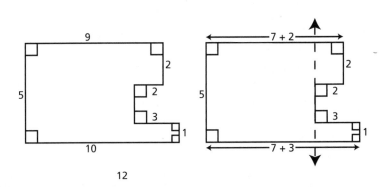

Continue lines to form rectangles within the polygon.

Find the area of each rectangle.

$7 \times 5 = 35$ $2 \times 2 = 4$
$3 \times 1 = 3$
Find the sum of all the areas.

$A = 35 + 4 + 3$
$A = 42$ square units

Use the formula to find the area of these polygons.

1 Area of each rectangle:

_____

_____

_____

2 A = _____ + _____ + _____

A = _____

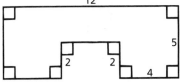

3 Area of each rectangle:

_____

_____

_____

4 A = _____ + _____ + _____

A = _____

Geometry © 2004 Creative Teaching Press

Formulas for Areas

Match the polygon with one or more formulas that can be used to find its area. Then find the area of the polygon.

A. $A = \dfrac{1}{2}bh$ **B.** $A = \dfrac{1}{2}(b_1 + b_2)h$ **C.** $A = bh$ *(or lw)*   **D.** $A = s^2$

1 Formula:_____

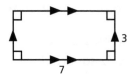

Area: _____

2 Formula:_____

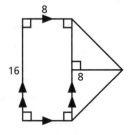

Area: _____

3 Formula:_____

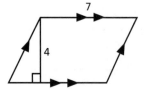

Area: _____

4 Formula:_____

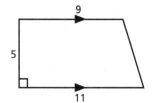

Area: _____

5 Formula:_____

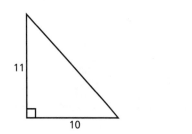

Area: _____

6 Formula:_____

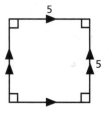

Area: _____

Geometry © 2004 Creative Teaching Press

Name _____ Date _____

Find Areas

Find the area for each polygon using the appropriate formula.

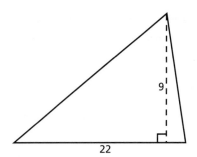

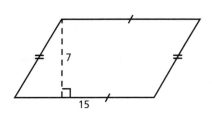

1 Formula used: _____

Area:_____

2 Formula used: _____

Area:_____

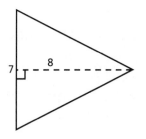

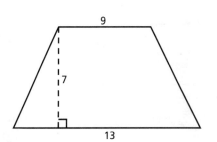

3 Formula used: _____

Area:_____

4 Formula used: _____

Area:_____

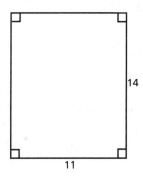

5 Formula used: _____

Area:_____

6 Formula used: _____

Area:_____

Geometry © 2004 Creative Teaching Press

Words for Circles

There are precise mathematical definitions for figures of circles and their parts. Match each term in the word box to its definition.

circle	interior	exterior	semicircle	diameter
radius	center	congruent	concentric	tangent
chord	secant	arc		

1 _____ This describes two or more circles on the same plane that share the same center.

2 _____ This is a line that intersects a circle in two points.

3 _____ This is a geometric figure in which all points in a plane are equidistant from a given point.

4 _____ This is a chord that contains the center of a circle.

5 _____ This describes coplanar circles that intersect in exactly one point.

6 _____ This is a continuous part of a circle. Two letters are used to name a minor one; three are used to name a major one.

7 _____ This is a segment with endpoints on a circle.

8 _____ This is a segment or distance from the center of a circle to a point on the circle.

9 _____ This describes the area inside the boundaries of a circle.

10 _____ This is an arc with endpoints that are the same as those of the diameter of the circle.

11 _____ This describes the area outside the boundaries of a circle.

12 _____ Circles are described as this if and only if they have equal radii.

13 _____ The point that is equidistant from all points on a circle.

Geometry © 2004 Creative Teaching Press

Name _____ Date _____

Label Circle Parts

Match each term in the word box to its description. Then use the numbers to label the corresponding part on the diagrams. Some terms are used more than once.

center	diameter	radius	chord	secant	semicircles
interior	exterior	tangent	major arc	minor arc	

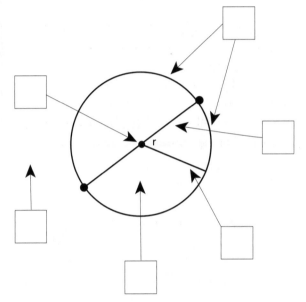 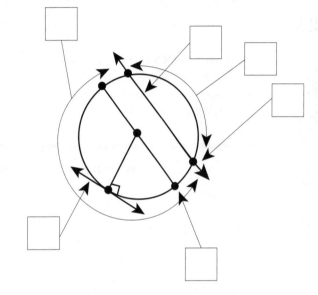

❶ _____ Two congruent half circles with endpoints that are intersected by the diameter.

❷ _____ All points on the circle are equidistant from this.

❸ _____ A segment that connects two points on the circle and contains the center of the circle.

❹ _____ The distance from the center to a point on the circle.

❺ _____ The area outside the circle's boundary.

❻ _____ The area inside the circle's boundary.

❼ _____ A chord that passes through the center of the circle.

❽ _____ A line segment with endpoints that are points on the circle.

❾ _____ A line that intersects the circle at one point.

❿ _____ A line that intersects the circle at two points.

⓫ _____ Part of a circle that is less than 180°.

⓬ _____ Part of a circle that is more than 180°.

Geometry © 2004 Creative Teaching Press

Congruent and Concentric Circles

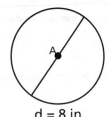 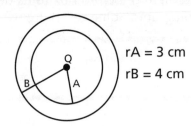

d = 8 in   r = 4 in

Congruent circles are those that have equal radii (plural for *radius*). Since the diameter is twice the radius, these two circles are congruent.

rA = 3 cm
rB = 4 cm

Concentric circles are coplanar circles that share the same center but not the same radius.
Both circles share Point Q. The radius of Circle A = 3 cm. The radius of Circle B = 4 cm. They are concentric.

Label each pair of circles as **congruent** or **concentric**. Then tell the radius of each circle.

 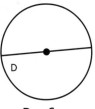

C = 3 cm   D = 6 cm

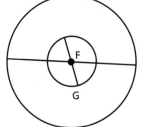

F = 2 ft
G = 11 ft

① _____

radius of _____ = _____

radius of _____ = _____

② _____

radius of _____ = _____

radius of _____ = _____

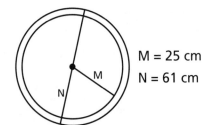

M = 25 cm
N = 61 cm

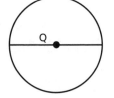

Q = 11 in   R = 5.5 in

③ _____

radius of _____ = _____

radius of _____ = _____

④ _____

radius of _____ = _____

radius of _____ = _____

Geometry © 2004 Creative Teaching Press

Tangent Circles

Coplanar circles that intersect in exactly one point are called **tangent circles.** They can be internally or externally tangent.

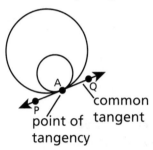

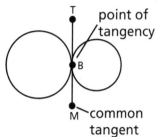

Internally tangent circles have one circle inside the other. Both share Point A on $\overleftrightarrow{PQ}$. $\overleftrightarrow{PQ}$ is also known as a common tangent.

Externally tangent circles are outside of one another but share exactly one common point. Both share Point B on $\overline{TM}$. $\overline{TM}$ is also known as a common tangent.

For each diagram, write **internally tangent** or **externally tangent** to identify the circles. Then name the point of tangency and the common tangent.

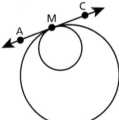

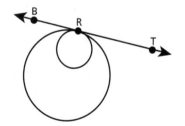

1 _____

2 Point of tangency: _____

3 Common tangent: _____

4 _____

5 Point of tangency: _____

6 Common tangent: _____

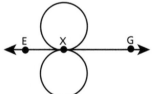

7 _____

8 Point of tangency: _____

9 Common tangent: _____

10 _____

11 Point of tangency: _____

12 Common tangent: _____

Geometry © 2004 Creative Teaching Press

Inscribed and Circumscribed Circles

Circles can be placed inside or outside a polygon.

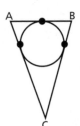
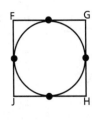
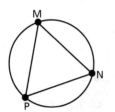
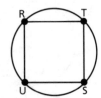

A circle is **inscribed** in a polygon if each side of the polygon is tangent to the circle at exactly one point.

A circle is **circumscribed** about a polygon if each vertex of the polygon lies on the circle.

For each diagram, write **inscribed** or **circumscribed**. Name the points of tangency and tell how many for each polygon.

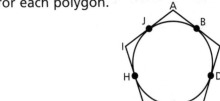

1 _____

2 Points of tangency: _____

_____

3 Number of tangent points: _____

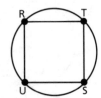

4 _____

5 Points of tangency: _____

_____

6 Number of tangent points: _____

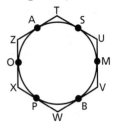

7 _____

8 Points of tangency: _____

_____

9 Number of tangent points: _____

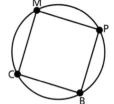

10 _____

11 Points of tangency: _____

_____

12 Number of tangent points: _____

Geometry © 2004 Creative Teaching Press

Name _____ Date _____

Diameters and Radii

The **diameter** is a segment connecting two points and contains the center of the circle.
d = 5 cm

The **radius** is the distance from the center to a point on the circle. The measure of a radius is half that of the diameter.
r = 2.5 cm

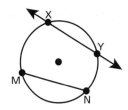

A **secant** is a line that intersects the circle at two points. A chord is a segment that intersects at two points.
$\overleftrightarrow{XY}$ is a secant.
$\overline{MN}$ is a chord.

Determine the diameter and radius of each diagram. Then name the secant and the chord.

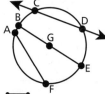

$\overleftrightarrow{BE}$ = 7.5 cm

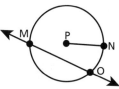

$\overleftrightarrow{PN}$ = 6 cm

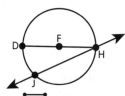

$\overleftrightarrow{DF}$ = 7 cm

1 radius = _____

   diameter = _____

   secant: _____

   chord: _____

2 radius = _____

   diameter = _____

   secant: _____

   chord: _____

3 radius = _____

   diameter = _____

   secant: _____

   chord: _____

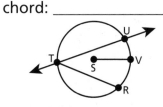

$\overleftrightarrow{SV}$ = 3.75 cm

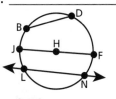

$\overleftrightarrow{JF}$ = 11 cm

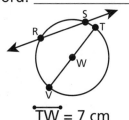

$\overleftrightarrow{TW}$ = 7 cm

4 radius = _____

   diameter = _____

   secant: _____

   chord: _____

5 radius = _____

   diameter = _____

   secant: _____

   chord: _____

6 radius = _____

   diameter = _____

   secant: _____

   chord: _____

Geometry © 2004 Creative Teaching Press

Name _____ Date _____

Area of a Circle

The area of a circle is π times the square of the radius.

Area of a Circle

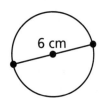

6 cm

$A = \pi r^2$

The radius is half of the diameter.
r = 3 cm

Square the radius.
$3^2 = 9$

Multiply by π.
A = 3.14 (9)
A = 28.26 cm

Find the radius for each circle. Then use the formula to find the area of each circle. Remember that π = 3.14.

9 cm

12 cm

1 Radius = _____

Use formula: _____

_____

_____

Area = _____

2 Radius = _____

Use formula: _____

_____

_____

Area = _____

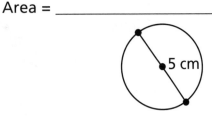

5 cm

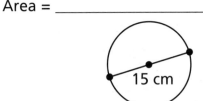

15 cm

3 Radius = _____

Use formula: _____

_____

_____

Area = _____

4 Radius = _____

Use formula: _____

_____

_____

Area = _____

Geometry © 2004 Creative Teaching Press

Name _____ Date _____

Arcs and Angles

An angle whose vertex is the center of a circle is a central angle of a circle. The sides of the angle are radii of the circle. Arcs split the circumference of the circle.

If the endpoints of an arc are the endpoints of a diameter, then the arc is a semicircle.

$\overset{\frown}{QM}$ is a semicircle.

∠MPN is a central angle.

$\overset{\frown}{MN}$ is a minor arc. It is less than 180°.

$\overset{\frown}{NQM}$ is a major arc. It is more than 180°. Major arcs are written with a third point so they are not confused with minor arcs.

Use the diagrams to write the names of the parts for each circle.

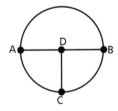

1 Name a major arc. _____

2 Name a minor arc. _____

3 Name a central angle. _____

4 Name a semicircle. _____

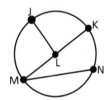

5 Name a major arc. _____

6 Name a minor arc. _____

7 Name a central angle. _____

8 Name a semicircle. _____

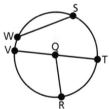

9 Name a major arc. _____

10 Name a minor arc. _____

11 Name a central angle. _____

12 Name a semicircle. _____

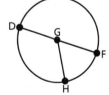

13 Name a major arc. _____

14 Name a minor arc. _____

15 Name a central angle. _____

16 Name a semicircle. _____

Geometry © 2004 Creative Teaching Press

Reflections

A **reflection** is the image of a figure that has been flipped over a line of reflection. The reflection image is called a mirror image or a flip.

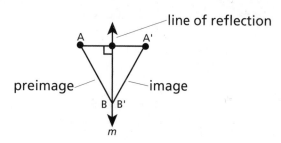

m is the line of reflection.

The preimage is the image before the reflection.

The image is the reflection. Point A corresponds to Point A'. Point B corresponds to Point B'.

Write **reflection** or **no reflection** to classify each diagram.

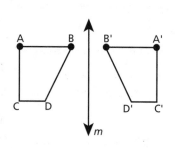

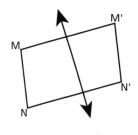

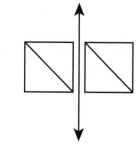

❶ _____ **❷** _____ **❸** _____ **❹** _____

Each of these diagrams shows a reflection. Draw lines of reflection for each figure.

❺  **❻** H **❼**   **❽** B

Geometry © 2004 Creative Teaching Press

Line Symmetry

A figure has **line symmetry** if it can be folded along a line so that it has two parts that are reflections of each other. The reflection line is called the line of symmetry. A figure can have no lines of symmetry, one line of symmetry, or more than one line of symmetry.

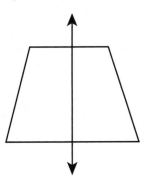

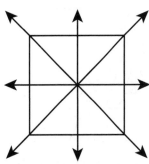

This figure has one line of symmetry. Both sides are reflections of the other

This figure has four lines of symmetry. On each line, the figure can be folded into two parts that are reflections of one another.

For each figure, write the quantity and draw the lines of symmetry. If a figure has no lines of symmetry, write **none**. If it has infinite lines of symmetry, only draw a few and write **infinite**.

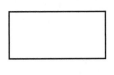

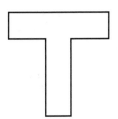

 ❶ _____

 ❷ _____

 ❸ _____

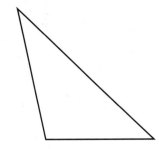

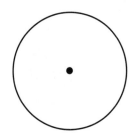

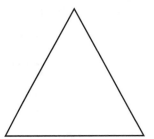

 ❹ _____

❺ _____

❻ _____

Geometry © 2004 Creative Teaching Press

Practice with Symmetry

For each figure, write the quantity and draw the lines of symmetry. If a figure has no lines of symmetry, write **none**. If it has infinite lines of symmetry, only draw a few and write **infinite**.

1 _____   **2** _____   **3** _____

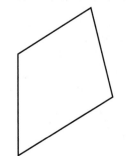

4 _____   **5** _____   **6** _____

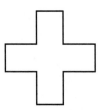

7 _____   **8** _____   **9** _____

Geometry © 2004 Creative Teaching Press

Name _____ Date _____

Rotations

A **rotation** is a transformation that is also called a turn. A rotation turns a figure about a point. This point is called the center of rotation. A rotation can be described by its direction, its amount, and its center of rotation.

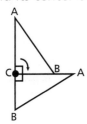

△ABC turns clockwise.
It turns $\frac{1}{4}$ turn.
It pivots on Point C, a part of the figure.

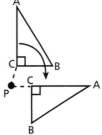

Here △ABC turns clockwise.
It turns $\frac{1}{4}$ turn.
It pivots on Point P, which is outside the figure.

For each rotation, tell the direction, the amount, and the center of rotation. Draw in the center of rotation if necessary. Draw an arrow to show the direction of rotation.

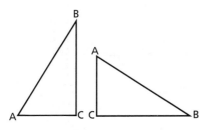

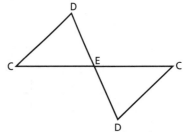

1 Direction: _____

2 Amount: _____

3 Center of rotation: _____

4 Direction: _____

5 Amount: _____

6 Center of rotation: _____

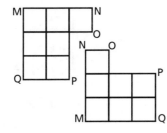

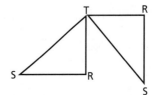

7 Direction: _____

8 Amount: _____

9 Center of rotation: _____

10 Direction: _____

11 Amount: _____

12 Center of rotation: _____

Geometry © 2004 Creative Teaching Press

Rotational Symmetry

A figure has **rotational symmetry** if it maps onto itself by a rotation of 180° or less. Mapping a figure onto itself means that the transformation moves each point so that it overlaps another point on the figure.

 → →

As the figure turns, it maps onto itself. It has rotational symmetry.

For each figure, draw its lines of symmetry. Then write **yes** or **no** to tell whether the figure has rotational symmetry.

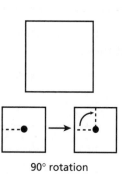

90° rotation

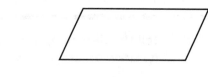

90° rotation   180° rotation

1 _____   **2** _____

60° rotation

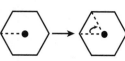

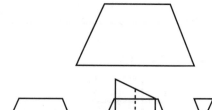

90° rotation   180° rotation

3 _____   **4** _____

5 What observations can you make about figures with rotational symmetry?

_____

Geometry © 2004 Creative Teaching Press

Name _____ Date _____

Translations

A **translation** is also called a slide or a glide. When a figure is translated, the line segments of the figure and its translated image are congruent and parallel.

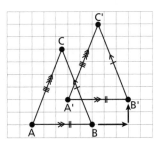

$\triangle$ABC slides over and up. The arrows show the direction of translation of the figure.

$\overline{AB}$ is parallel and congruent to $\overline{A'B'}$.
$\overline{BC}$ is parallel and congruent to $\overline{B'C'}$.
$\overline{AC}$ is parallel and congruent to $\overline{A'C'}$.

For each figure, draw arrows to show the direction of translation. Then name the congruent and parallel parts. Explain the translation that occurred.

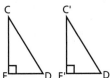

1 _____

_____

_____

What was the translation? _____

2 _____

_____

_____

What was the translation? _____

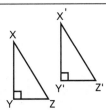

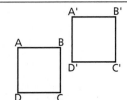

3 _____

_____

_____

What was the translation? _____

4 _____

_____

_____

What was the translation? _____

Geometry © 2004 Creative Teaching Press

Name _____ Date _____

Practice with Transformations

For each transformation, write the kind of transformation that occurred and draw arrows or lines of reflection to show how the image changed.

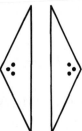

1 Transformation:_____

2 How did it change? _____

_____

3 Transformation:_____

4 How did it change? _____

_____

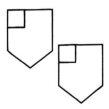

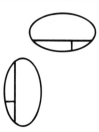

5 Transformation:_____

6 How did it change? _____

_____

7 Transformation:_____

8 How did it change? _____

_____

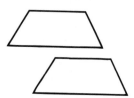

9 Transformation:_____

10 How did it change? _____

_____

11 Transformation:_____

12 How did it change? _____

_____

Geometry © 2004 Creative Teaching Press

Answer Key

Geometric Vocabulary (page 5)

1. point

2. line

3. collinear

4. plane

5. coplanar

6. line segment

7. ray

8. angle

9. point

10. angle

11. line

12. ray

13. collinear

14. coplanar

15. line segment

16. plane

Identify Lines, Line Segments, and Rays (page 6)

Note: The two points that name a line or segment can be written in either order.

1. $\overleftrightarrow{PQ}$ or $\overleftrightarrow{QP}$, $\overleftrightarrow{PR}$ or $\overleftrightarrow{RP}$, $\overleftrightarrow{QR}$ or $\overleftrightarrow{RQ}$

2. $\overrightarrow{PR}$, $\overrightarrow{PQ}$, $\overrightarrow{QP}$, $\overrightarrow{QR}$, $\overrightarrow{RQ}$, $\overrightarrow{RP}$

3. $\overline{PQ}$ or $\overline{QP}$, $\overline{QR}$ or $\overline{RQ}$, $\overline{RP}$ or $\overline{PR}$

4. $\overleftrightarrow{MN}$ or $\overleftrightarrow{NM}$, $\overleftrightarrow{NO}$ or $\overleftrightarrow{ON}$, $\overleftrightarrow{MO}$ or $\overleftrightarrow{OM}$

5. $\overrightarrow{MN}$, $\overrightarrow{MO}$, $\overrightarrow{NM}$, $\overrightarrow{NO}$, $\overrightarrow{ON}$, $\overrightarrow{OM}$

6. $\overline{MN}$ or $\overline{NM}$, $\overline{NO}$ or $\overline{ON}$, $\overline{MO}$ or $\overline{OM}$

Name Lines, Line Segments, and Rays (page 7)

1. $\overrightarrow{JK}$ $\overleftrightarrow{KL}$ $\overrightarrow{JL}$

2. $\overleftrightarrow{RU}$ $\overleftrightarrow{UV}$ $\overrightarrow{RV}$ $\overleftrightarrow{VU}$

3. $\overline{EF}$ $\overline{XF}$ $\overline{EX}$ $\overline{FE}$

Find Lengths of Line Segments (page 8)

1. 5

2. 1

3. 2

4. 3

5. 6

6. 8

7. 11

8. 3

9. 8

10. 6

11. 5

12. 11

13. $\overline{CE}$ or $\overline{EC}$

14. $\overline{BD}$ or $\overline{DB}$

15. $\overline{AB}$ or $\overline{BA}$

16. $\overline{DE}$ or $\overline{ED}$

Collinear Points and Line Segments (page 9)

1. false

2. true

3. true

4. false

5. false

6. true

7. false

8. true

9. false

10. true

11. false

12. true

Identify Objects in a Plane (page 10)

1. Point M, N, R, or Q

2. Point S or Point X

3. $\overrightarrow{NP}$ or $\overrightarrow{PN}$, $\overrightarrow{PQ}$ or $\overrightarrow{QP}$, $\overrightarrow{RQ}$ or $\overrightarrow{QR}$

4. $\overleftrightarrow{XS}$ or $\overleftrightarrow{SX}$, $\overleftrightarrow{RS}$ or $\overleftrightarrow{SR}$, $\overleftrightarrow{QX}$ or $\overleftrightarrow{XQ}$

5. Point Q, Point P, Point N, and Point R

6. Point B or Point A

7. $\overrightarrow{FG}$ or $\overrightarrow{GF}$, $\overrightarrow{GD}$ or $\overrightarrow{DG}$, $\overrightarrow{DE}$ or $\overrightarrow{ED}$

8. $\overleftrightarrow{AD}$ or $\overleftrightarrow{DA}$, $\overrightarrow{GA}$ or $\overrightarrow{AG}$

9. Point G, Point F, Point E, Point L, or Point D

10. $\overline{EF}$ or $\overline{FE}$, $\overline{FG}$ or $\overline{GF}$, $\overline{GD}$ or $\overline{DG}$

Line and Segment Relationships (page 11)

1. intersecting; line A intersects line B

2. concurrent; lines A, B, C, and D are concurrent

3. perpendicular; $\overleftrightarrow{AB} \perp \overleftrightarrow{CD}$

4. parallel; $\overleftrightarrow{AB} \parallel \overleftrightarrow{CD}$

5. skew; $\overleftrightarrow{AB}$ and $\overleftrightarrow{CD}$ are skew

Intersecting and Parallel Lines in Planes (page 12)

1. $\overleftrightarrow{CE}$ or $\overleftrightarrow{EC}$, $\overleftrightarrow{DF}$ or $\overleftrightarrow{FD}$, $\overleftrightarrow{CB}$ or $\overleftrightarrow{BC}$, $\overleftrightarrow{DA}$ or $\overleftrightarrow{AD}$

2. $\overleftrightarrow{AB}$ or $\overleftrightarrow{BA}$, $\overleftrightarrow{EF}$ or $\overleftrightarrow{FE}$

3. $\overleftrightarrow{CE}$ or $\overleftrightarrow{EC}$, $\overleftrightarrow{FD}$ or $\overleftrightarrow{DF}$

4. $\overleftrightarrow{CD}$ or $\overleftrightarrow{DC}$

5. false; they are parallel

6. true

7. false; they are parallel

8. false; they are not on the same plane

Identify Intersecting and Parallel Segments (page 13)

1. false

2. true

3. true

4. false

5. true

6. false

7. true

8. true

9. false

10. false

Find Geometric Answers (page 14)

1. $\overleftrightarrow{QP}$, $\overleftrightarrow{PR}$ or $\overleftrightarrow{RP}$, $\overleftrightarrow{RQ}$ or $\overleftrightarrow{QR}$

2. $\overleftrightarrow{PO}$ or $\overleftrightarrow{OP}$, $\overleftrightarrow{TR}$ or $\overleftrightarrow{RT}$, $\overleftrightarrow{PQ}$ or $\overleftrightarrow{QP}$, $\overleftrightarrow{PR}$ or $\overleftrightarrow{RP}$, $\overleftrightarrow{QR}$ or $\overleftrightarrow{RQ}$, $\overleftrightarrow{XY}$ or $\overleftrightarrow{YX}$

3. $\overleftrightarrow{RT}$ or $\overleftrightarrow{TR}$, $\overleftrightarrow{XY}$ or $\overleftrightarrow{YX}$

4. $\overrightarrow{QP}$, $\overrightarrow{PR}$, $\overrightarrow{RQ}$, $\overrightarrow{RT}$, $\overrightarrow{XY}$

5. $\overrightarrow{OP}$, $\overrightarrow{RT}$, $\overrightarrow{XY}$

6. Possible answers: $\overrightarrow{PR}$ and $\overrightarrow{PO}$, $\overrightarrow{PR}$ and $\overleftrightarrow{RT}$

7. a ray because it only has one endpoint

8. yes, because they both name points on the same line

9. yes, because they are on the same plane but do not intersect

10. yes, if both continued on they would eventually cross each other

Congruence of Segments (page 15)

1. true
2. true
3. false
4. false
5. true
6. true
7. true
8. true
9. ⟨JK⟩ = 4; $\overline{LM}$ = 2; ⟨MN⟩ = 3; ⟨PQ⟩ = 4
10. ⟨NO⟩ = 2; $\overline{MN}$ = 3; ⟨QR⟩ = 2; ⟨KL⟩ = 2
11. ⟨KL⟩ = 2; $\overline{OP}$ = 1; ⟨QR⟩ = 2; $\overline{PQ}$ = 4
12. $\overline{MN}$ = 3; $\overline{ON}$ = 2; ⟨QP⟩ = 4; ⟨KJ⟩ = 4

Add to Find Congruent Segments (page 16)

1. false
2. false
3. false
4. false
5. true
6. true
7. true
8. false
9. ⟨JK⟩ = 4; $\overline{JM}$ = 8; ⟨MK⟩ = 4; $\overline{RP}$ = 6
10. ⟨LO⟩ = 7; $\overline{MN}$ = 3; $\overline{PR}$ = 6; ⟨KN⟩ = 7
11. $\overline{MJ}$ = 8; ⟨OL⟩ = 7; ⟨QN⟩ = 7; $\overline{PR}$ = 6
12. $\overline{MP}$ = 6; ⟨JM⟩ = 8; $\overline{KN}$ = 7; ⟨LP⟩ = 8

Subtract to Find Segment Lengths (page 17)

1. 9, 6, 15
2. 20, 39, 8.5
3. 30, 78, 48
4. 44, 16, 26
5. 18
6. 21
7. 4
8. 8
9. 21
10. 18

(continued top right)

11. 39
12. 12
13. 33

Find Congruent Line Segments (page 18)

1. 18
2. 40
3. 44
4. 52
5. 4
6. 12
7. 34
8. 52
9. 26
10. 34
11. 74
12. 8
13. 30
14. 48
15. 22
16. 40
17. true
18. false
19. false
20. true
21. false
22. true
23. false
24. false
25. false
26. true

Geometry and Algebraic Equations (page 19)

1. x = 19
2. x = 9
3. x = 8
4. x = 5

Angle Vocabulary (page 20)

1. vertex
2. linear
3. adjacent
4. obtuse
5. side
6. right
7. vertical
8. complementary
9. supplementary
10. straight
11. bisector
12. acute

Name and Label Angles (page 21)

1. Point K
2. $\overrightarrow{KJ}$ and $\overrightarrow{KL}$
3. ∠2, ∠K, ∠JKL, ∠LKJ
4. Point O
5. $\overrightarrow{OB}$ and $\overrightarrow{OX}$
6. ∠3, ∠O, ∠XOB, ∠BOX
7. Point R
8. $\overrightarrow{RP}$ and $\overrightarrow{RQ}$
9. ∠4, ∠R, ∠PRQ, ∠QRP

Name and Label Connected Angles (page 22)

1. Point E
2. $\overrightarrow{ES}$
3. ∠1, ∠RES, ∠SER
4. ∠2, ∠SET, ∠TES
5. ∠RET, ∠TER
6. Point U
7. $\overrightarrow{UN}$
8. ∠1, ∠MUN, ∠NUM, ∠2, ∠NUR, ∠RUN, ∠MUR, ∠RUM

Classify Angles (page 23)

1. obtuse; ∠I, ∠GIV, ∠VIG
2. right; ∠D, ∠ADF, ∠FDA
3. acute; ∠O, ∠XOT, ∠TOX
4. obtuse; ∠N, ∠WNF, ∠FNW
5. acute; ∠B, ∠ABC, ∠CBA
6. straight; ∠K, ∠JKL, ∠LKJ

Identify Congruent Angles (page 24)

Note: For all angles and lines, letters can be written in reverse order.

1. ∠IFJ, ∠JFK, ∠GFH
2. ∠GFK, ∠IFK, ∠IFG, ∠JFH
3. ∠HFI
4. ∠IFJ ≅ ∠JFK
5. $\overleftrightarrow{JG}$
6. ∠TSR, ∠RSU, ∠TSU, or ∠USV
7. ∠TSV
8. ∠RSV
9. ∠RSU ≅ ∠USV
10. $\overrightarrow{SU}$ (can only be written in this order)

Add and Subtract to Find Angle Measures (page 25)

1. 23°
2. 45°
3. 87°
4. 135°
5. 180°
6. m∠DAB − m∠MAB = 87 − 23 = 64°
7. m∠KAB − m∠GAB = 180 − 135 = 45°
8. m∠KAB − m∠DAB = 180 − 87 = 93°
9. m∠GAB − m∠LAB = 135 − 45 = 90°
10. m∠GAB − m∠MAB = 135 − 23 = 112°
11. m∠KAB − m∠LAB = 180 − 45 = 135°
12. ∠GAB ≅ ∠LAK, ∠LAB ≅ ∠KAG

Calculate Angle Measures (page 26)

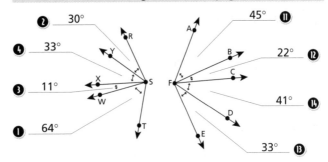

5. 63°
6. 138°
7. 75°
8. 44°
9. 108°
10. 74°
15. 67°
16. 63°
17. 108°
18. 96°
19. 141°
20. 74°

Complementary and Perpendicular Angles (page 27)

1. 55°, 35°, adjacent
2. 30°, 60°, nonadjacent
3. 65°, 25°, nonadjacent
4. 42°, 48°, adjacent
5. 72°, 18°, adjacent
6. 38°, 52°, adjacent

Supplementary and Linear Angles (page 28)

1. 25°, 155°, linear
2. 45°, 135°
3. 73°, 107°
4. 150°, 30°
5. 55°, 125°, linear
6. 153°, 27°, linear

Find Values of Angles (page 29)

1. 148°; Both total 180°
2. 78°; Both total 180°
3. 82°; Both adjacent angles total 180°
4. 49°; Both total 90°
5. 137°; Both total 180°
6. 87°; Both adjacent angles total 180°
7. 123°; Both adjacent angles total 180°
8. 69°; Both total 180°

9. 74°; Both total 90°

10. 162°; Both total 180°

Vertical Angles (page 30)

1. 142°, 38°

∠XAY and ∠TAZ

∠XAT and ∠YAZ

2. 118°, 62°

∠PQN and ∠JQM

∠JQP and ∠MQN

3. 108°, 72°

∠JKL and ∠UKM

∠JKU and ∠LKM

4. 90°, 90°

∠RSU and ∠FSM

∠RSF and ∠USM

Identify Types of Angles (page 31)

1. ∠6 and ∠4; ∠3 and ∠5

2. ∠1 and ∠2

3. ∠6 and ∠4

4. ∠3 and ∠4; ∠5 and ∠6; ∠1 and ∠2

5. ∠DGJ and ∠FGH; ∠DGF and ∠HGJ

6. ∠CFB and ∠BFJ; ∠FGD and ∠DGJ; ∠FGH and ∠HGJ

7. ∠DGF and ∠HGJ

8. ∠FGD and ∠DGJ; ∠FGH and ∠HGJ; ∠FGD and ∠FGH; ∠DGJ and ∠HGJ

Find Missing Angle Measurements (page 32)

1. 58° **2.** 27°

3. 79° **4.** 112°

5. 69° **6.** 93°

7. 113° **8.** 48°

9. 136° **10.** 56°, 124°

11. 40°, 140° **12.** 48°, 42°

Determine Missing Angles (page 33)

1. 29°, 29° **2.** 59°, 93°

3. 39°, 121°, 59° **4.** 79°, 63

Angles and Algebra (page 34)

1. 4x
8x
8x + 4x = 180°
x = 15
60°, 120°

2. (14x + 7)
(20x + 15)
(14x + 7) + (20x + 15) = 90°
x = 2
35°, 55°

More Applications of Algebra (page 35)

1. 3x + 15
2x
3x + 15 + 2x = 90°
x = 15
60°, 30°

2. x
4x − 20
x + 4x − 20 = 180°
x = 40
40°, 140°

3. 5x − 27
4x
5x − 27 + 4x =180
x = 23
88°, 92°
92°, 88°

Vocabulary to Describe Intersections (page 36)

1. intersect **2.** perpendicular

3. corresponding **4.** transversal

5. exterior **6.** parallel

7. alternate **8.** bisector

9. same side **10.** interior

Perpendicular Lines (page 37)

1. $\overleftrightarrow{AB}$, $\overleftrightarrow{PQ}$, $\overleftrightarrow{ST}$

2.

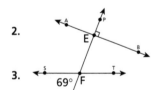

3.

4. right

5. They are perpendicular.

6. 69°, 111°, 111°

7. They intersect.

Parallel Lines (page 38)

1. Possible answers: $\overleftrightarrow{PJ}$ and $\overleftrightarrow{MR}$, $\overleftrightarrow{PJ}$ and $\overleftrightarrow{FX}$, $\overleftrightarrow{PJ}$ and $\overleftrightarrow{QD}$, $\overleftrightarrow{FX}$ and $\overleftrightarrow{MR}$, or $\overleftrightarrow{QD}$ and $\overleftrightarrow{MR}$
2. $\overleftrightarrow{QD}$ and $\overleftrightarrow{FX}$
3. $\overleftrightarrow{PJ}$ and $\overleftrightarrow{MR}$
4. Possible answers: $\overleftrightarrow{TO}$ and $\overleftrightarrow{RD}$, $\overleftrightarrow{TO}$ and $\overleftrightarrow{CA}$, $\overleftrightarrow{CA}$ and $\overleftrightarrow{RD}$, $\overleftrightarrow{CA}$ and $\overleftrightarrow{PS}$, $\overleftrightarrow{PS}$ and $\overleftrightarrow{TO}$
5. $\overleftrightarrow{SP}$ and $\overleftrightarrow{RD}$
6. $\overleftrightarrow{TO}$ and $\overleftrightarrow{CA}$

Transversals and Corresponding Angles (page 39)

1. $\overleftrightarrow{AB}$
2. $\overleftrightarrow{CD}$ and $\overleftrightarrow{FE}$
3. $\angle 2$ and $\angle 6$, $\angle 4$ and $\angle 8$, $\angle 1$ and $\angle 5$, $\angle 3$ and $\angle 7$
4. $\overleftrightarrow{RS}$
5. $\overleftrightarrow{LM}$ and $\overleftrightarrow{FG}$
6. $\angle 1$ and $\angle 5$, $\angle 4$ and $\angle 8$, $\angle 2$ and $\angle 6$, $\angle 3$ and $\angle 7$

Interior Angles (page 40)

1. $\overleftrightarrow{RS}$
2. $\overleftrightarrow{XY}$, $\overleftrightarrow{MO}$
3. $\angle 3$ and $\angle 5$, $\angle 4$ and $\angle 6$
4. $\angle 3$ and $\angle 6$, $\angle 5$ and $\angle 4$
5. $\overleftrightarrow{LS}$
6. $\overleftrightarrow{AB}$, $\overleftrightarrow{MN}$
7. $\angle 4$ and $\angle 5$, $\angle 3$ and $\angle 6$
8. $\angle 4$ and $\angle 6$, $\angle 3$ and $\angle 5$

Transversals and Exterior Angles (page 41)

1. $\overleftrightarrow{KM}$
2. $\overleftrightarrow{EH}$, $\overleftrightarrow{NP}$
3. $\angle 1$ and $\angle 7$, $\angle 2$ and $\angle 8$
4. $\angle 1$ and $\angle 8$, $\angle 2$, and $\angle 7$
5. $\overleftrightarrow{QK}$
6. $\overleftrightarrow{CD}$, $\overleftrightarrow{PT}$
7. $\angle 1$ and $\angle 8$, $\angle 2$ and $\angle 7$
8. $\angle 1$ and $\angle 7$, $\angle 2$ and $\angle 8$

Identify Transversal Angles (page 42)

1. $\angle PNM$ and $\angle RQT$, $\angle PNO$ and $\angle SQT$
2. $\angle MNT$ and $\angle PQR$, $\angle PQS$ and $\angle ONT$
3. $\angle MNT$ and $\angle PQS$, $\angle ONT$ and $\angle PQR$
4. $\angle PNM$ and $\angle SQT$, $\angle PNO$ and $\angle RQT$
5. Possible answers: $\angle ABD$ and $\angle FGD$, $\angle DBC$ and $\angle DGH$, $\angle ABE$ and $\angle FGE$, $\angle EBC$ and $\angle EGH$
6. $\angle ABD$ and $\angle EGH$, $\angle DBC$ and $\angle FGE$
7. $\angle ABE$ and $\angle DGH$, $\angle FGD$ and $\angle CBE$
8. $\angle ABD$ and $\angle FGE$, $\angle DBC$ and $\angle EGH$
9. true
10. false
11. false
12. true

Congruent Transversal Angle Measures (page 43)

1. 80°	2. 80°
3. 100°	4. 100°
5. 80°	6. 100°
7. 46°	8. 134°
9. 46°	10. 46°
11. 134°	12. 134°

Supplementary Transversal Angle Measures (page 44)

1. 107°	2. 73°
3. 107°	4. 75°
5. 105°	6. 75°
7. 70°	8. 110°
9. 70°	10. 56°
11. 124°	12. 124°

More Transversal Angles (page 45)

1. 121°	2. 121°
3. 59°	4. 77°
5. 103°	6. 77°
7. 66°	8. 114°
9. 114°	10. 66°
11. 114°	12. 114°
13. 66°	

Midpoint Bisectors (page 46)

1. $-4 + 6$
midpoint coordinate = 1

2. $-3 + 10$
midpoint coordinate = 3.5

3. $-6 + 3$
midpoint coordinate = -1.5

4. $3 + 9$
midpoint coordinate = 6

5. $-11 + 1$
midpoint coordinate = -5

6. $-9 + 3$
midpoint coordinate = -3

7. $-2 + 8$
midpoint coordinate = 3

8. $-3 + 13$
midpoint coordinate = 5

9. $-1 + 11$
midpoint coordinate = 5

Angle Bisectors (page 47)

1. $\overrightarrow{JM}$

2. m∠KJM = m∠MJL

3. 60°, 30°

4. $\overrightarrow{PN}$

5. m∠MPN = m∠NPO

6. 24°, 24°

7. $\overrightarrow{SP}$

8. 56°, 28°

9. m∠TVW = m∠WVX

10. 16°, 32°

Algebra and Intersecting Lines (page 48)

1. $4x + 9x + 11 = 180°$
$x = 13$
128°
52°

2. $x + 5 + 6x = 180°$
$x = 25$
30°
150°

3. $x = 25$
90°
90°

4. $x = 16$
148°
32°

Algebra and Transversals (page 49)

1. $11x + 2 + 7x - 2 = 180°$
$x = 10$
68°
112°
112°
112°

2. $6x + 12 + 5x + 3 = 180°$
$8x - 5 + 4x + 5 = 180°$
$x = 15$
102°
78°
115°
65°

3. $11x + 4x = 180°$
$x = 12$
132°
48°
90°
90°

4. $4x + 10 + 3x - 5 = 180°$
$x = 25$
110°
70°
70°
110°

Triangles (page 50)

1. acute | **2.** triangle

3. altitude | **4.** obtuse

5. scalene | **6.** hypotenuse

7. equiangular | **8.** legs

9. right | **10.** base

11. isosceles | **12.** equilateral

Classify Triangles by Sides (page 51)

1. isosceles | **2.** scalene

3. isosceles | **4.** scalene

5. scalene | **6.** scalene

7. equilateral | **8.** isosceles

9. equilateral

Classify Triangles by Angles (page 52)

1. right | **2.** right

3. acute | **4.** right

5. acute | **6.** obtuse

7. equiangular | **8.** obtuse

9. equiangular

What Triangle Is It? (page 53)

1. right scalene
2. obtuse scalene
3. equiangular equilateral
4. acute scalene
5. right scalene
6. right scalene
7. obtuse scalene
8. acute isosceles
9. right scalene

Find Missing Triangle Measures (page 54)

1. 40°, 70°
2. 58°, 90°
3. 74°
4. 30°, 90°
5. 65°
6. 48°, 66°
7. 60°, 60°
8. 68°, 44°
9. 55°

Determine Triangle Measures (page 55)

1. 90° and 40°
2. 45°
3. 60° and 60°
4. 54°
5. 90° and 30°
6. 110°
7. 108°
8. 36°
9. 90° and 55°
10. 90° and 32°
11. 60° and 60°
12. 100°

Segments in Triangles (page 56)

1. JC
2. AC
3. $\overline{JC}$, $\overline{AC}$
4. $\overline{AJ}$, $\overline{AC}$
5. PN
6. MN
7. $\overline{PM}$, $\overline{PN}$
8. $\overline{NP}$, $\overline{NM}$
9. ∠T
10. ∠B
11. ∠T
12. ∠X

Segments in Isosceles Triangles (page 57)

1. ∠Z, ∠Y
2. $\overline{XZ}$, $\overline{XY}$
3. m∠Y = 50°
4. m $\overline{XZ}$ = 6 cm
5. ∠P, ∠Q
6. $\overline{NP}$, $\overline{NQ}$
7. m∠P = 76°
8. m $\overline{NP}$ = 11 cm

The Medians of a Triangle (page 58)

1. $\overline{SX}$, $\overline{NT}$, $\overline{BP}$
2. Point F
3. a side
4. no
5. yes
6. $\overline{JM}$, $\overline{CT}$, $\overline{AX}$
7. Point B
8. a median
9. yes
10. yes

The Perpendicular Bisectors of a Triangle (page 59)

1. $\overrightarrow{AQ}$, $\overrightarrow{PQ}$, $\overrightarrow{SQ}$
2. Point Q
3. a side
4. yes
5. $\overleftrightarrow{HK}$, $\overleftrightarrow{FH}$, $\overleftrightarrow{HQ}$
6. Point H
7. yes
8. yes, because an angle is bisected and creates two equal parts

The Angle Bisectors of a Triangle (page 60)

1. $\overrightarrow{BO}$, $\overrightarrow{TN}$, $\overrightarrow{XM}$
2. Point P
3. m∠BTN = 34°, m∠NTX = 34°
4. m∠BXM = 16°, m∠MXT = 16°
5. $\overrightarrow{JE}$, $\overrightarrow{CF}$, $\overrightarrow{AH}$
6. Point G
7. m∠ACF = 17.5°, m∠FCJ = 17.5°
8. m∠JCF = 27.5°, m∠FCA = 27.5°

The Altitude of a Triangle (page 61)

1. $\overline{BM}$, $\overline{LT}$, $\overline{KX}$
2. Point N
3. $\overline{LT}$
4. $\overline{MX}$, $\overline{MT}$, or $\overline{TX}$
5. $\overline{AS}$, $\overline{RX}$, $\overline{CT}$
6. Point X
7. $\overline{AS}$
8. $\overline{RX}$

Determine Concurrent Lines (page 62)

1. median
2. angle bisector
3. perpendicular bisector
4. perpendicular bisector
5. altitude
6. median
7. altitude
8. angle bisector
9. median

Quadrilaterals (page 63)

1. diagonals
2. parallelogram
3. rhombus
4. square
5. quadrilateral
6. opposite angles
7. trapezoid
8. rectangle
9. opposite sides
10. leg
11. trapezium
12. kite
13. base

Name Quadrilaterals (page 64)

1. quadrilateral, parallelogram
2. quadrilateral, parallelogram, rectangle, square
3. quadrilateral, parallelogram, rectangle
4. quadrilateral, trapezoid
5. quadrilateral, trapezium
6. quadrilateral, parallelogram, rhombus
7. quadrilateral
8. quadrilateral, parallelogram, rectangle
9. quadrilateral, trapezium, kite

Use Markings to Show Quadrilaterals (page 65)

1. opposite sides are congruent
   opposite sides are parallel
   all sides are perpendicular
   rectangle
2. one pair of opposite sides is parallel
   no sides or angles are congruent
   trapezoid
3. all angles are perpendicular
   opposite sides are parallel
   all four sides are congruent
   square
4. opposite sides are parallel
   all sides are congruent
   rhombus

Diagonals on Quadrilaterals (page 66)

1. $\overline{PR}$ and $\overline{QS}$
2. $\overline{PS}$ and $\overline{QR}$; $\overline{PQ}$ and $\overline{SR}$
3. $\angle QRS$ and $\angle SPQ$; $\angle PQR$ and $\angle RSP$

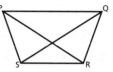

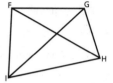

4. $\overline{GI}$ and $\overline{FH}$
5. $\overline{FI}$ and $\overline{GH}$; $\overline{FG}$ and $\overline{HI}$
6. $\angle IFG$ and $\angle GHI$; $\angle FIH$ and $\angle HGF$

Parallelograms (page 67)

1. $\overline{BC} \cong \overline{ED}$; $\overline{BE} \cong \overline{CD}$
2. $\angle C \cong \angle E$; $\angle B \cong \angle D$
3. $m\angle E$ and $m\angle D = 180°$; $m\angle C$ and $m\angle B = 180°$;
   $m\angle E$ and $m\angle B = 180°$; $m\angle C$ and $m\angle D = 180°$
4. $\overline{BF} \cong \overline{FD}$; $\overline{CF} \cong \overline{FE}$
5. $\overline{JK} \cong \overline{ML}$; $\overline{JM} \cong \overline{KL}$
6. $\angle J \cong \angle L$; $\angle K \cong \angle M$
7. $m\angle J$ and $m\angle K = 180°$; $m\angle M$ and $m\angle L = 180°$;
   $m\angle J$ and $m\angle M = 180°$; $m\angle K$ and $m\angle L = 180°$
8. $\overline{JN} \cong \overline{NL}$; $\overline{MN} \cong \overline{NK}$

Rectangles (page 68)

1. $\overline{RS} \cong \overline{TU}$; $\overline{RT} \cong \overline{SU}$
2. $\angle R \cong \angle S \cong \angle U \cong \angle T$
3. $\overline{RV} \cong \overline{VU}$
   $\overline{SV} \cong \overline{VT}$
4. $m\angle R = 90°$; $m\angle S = 90°$; $m\angle T = 90°$; $m\angle U = 90°$
5. $\overline{FI} \cong \overline{GH}$; $\overline{HI} \cong \overline{FG}$
6. $m\angle F$ and $m\angle G = 180°$; $m\angle G$ and $m\angle H = 180°$;
   $m\angle H$ and $m\angle I = 180°$; $m\angle I$ and $m\angle F = 180°$
7. $\overline{FK} \cong \overline{KH}$
   $\overline{GK} \cong \overline{KI}$
8. $m\angle I = 90°$; $m\angle F = 90°$; $m\angle G = 90°$; $m\angle H = 90°$

Rhombi (page 69)

1. $\overline{EG} \perp \overline{FH}$

2. $\overline{EG}$ bisects $\angle G$ and $\angle E$; $\overline{FH}$ bisects $\angle F$ and $\angle H$

3. $\overline{FG} \cong \overline{GH} \cong \overline{HE} \cong \overline{EF}$

4. $\angle F \cong \angle H$; $\angle E \cong \angle G$

5. $\overline{PM} \perp \overline{QN}$

6. $\overline{MP}$ bisects $\angle M$ and $\angle P$; $\overline{QN}$ bisects $\angle Q$ and $\angle N$

7. $\overline{MQ} \cong \overline{MN} \cong \overline{NP} \cong \overline{PQ}$

8. $\angle Q \cong \angle N$; $\angle P \cong \angle M$

Squares (page 70)

1. $\angle F$ or $\angle G$ or $\angle H$ or $\angle J$

2. $\angle F \cong \angle G \cong \angle H \cong \angle J$

3. $\overline{FG} \cong \overline{GH} \cong \overline{HJ} \cong \overline{JF}$

4. $\overline{FH} \perp \overline{GJ}$

5. $\overline{RT} \perp \overline{SW}$

6. $\overline{RT}$ bisects $\angle T$ and $\angle R$; $\overline{SW}$ bisects $\angle S$ and $\angle W$

7. $\overline{RS} \cong \overline{ST} \cong \overline{TW} \cong \overline{RW}$

8. $\angle R \cong \angle S \cong \angle T \cong \angle W$

Trapezoids (page 71)

1. $\overline{KN}$; $\overline{LM}$

2. $\overline{LK}$; $\overline{MN}$

3. $\angle K$ and $\angle N$; $\angle L$ and $\angle M$

4. Yes, because the legs and diagonals are congruent.

5. $\overline{SR}$; $\overline{PQ}$

6. $\overline{PS}$; $\overline{QR}$

7. $\angle S$ and $\angle R$; $\angle P$ and $\angle Q$

8. No, because $\overline{QR}$ is longer and the diagonals are not congruent.

Kites (page 72)

1. trapezium; It does not have perpendicular diagonals.

2. trapezium; It has no opposite congruent angles.

3. kite; It has opposite congruent angles and perpendicular diagonals.

4. kite; It has opposite congruent angles and perpendicular diagonals.

Define Polygons (page 73)

1. nonagon
2. heptagon
3. concave
4. interior angle
5. octagon
6. exterior angle
7. convex
8. regular
9. hexagon
10. pentagon
11. irregular
12. decagon
13. dodecagon

Identify Polygons (page 74)

1. polygon

2. not a polygon; It has two sides that do not meet.

3. not a polygon; It has a curve.

4. polygon

5. polygon

6. not a polygon; It has a curve.

Convex and Concave Polygons (page 75)

1. convex
2. convex
3. concave
4. convex
5. convex
6. concave

Identify by Sides (page 76)

1. quadrilateral
2. hexagon
3. triangle
4. dodecagon
5. octagon
6. nonagon
7. pentagon
8. heptagon
9. decagon

Diagonals of Polygons (page 77)

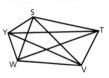
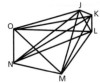

1. n = 6
   n − 1 = 5

2. n = 6
   n − 1 = 5

3. n = 4
   n − 1 = 3

4. n = 6
   n − 1 = 5

5. n = 8
   n − 1 = 7

6. n = 5
   n − 1 = 4

Regular and Irregular Polygons (page 78)

1. 6

2. irregular, hexagon

3. $\overline{AB} \cong \overline{BC} \cong \overline{FE} \cong \overline{DE}$; $\overline{CD} \cong \overline{AF}$

4. 7

5. regular, heptagon

6. $\overline{XM} \cong \overline{MT} \cong \overline{TS} \cong \overline{SV} \cong \overline{VW} \cong \overline{WN} \cong \overline{XN}$

7. 4

8. irregular, quadrilateral

9. $\overline{FB} \cong \overline{DC}$;
   $\overline{FD} \cong \overline{BC}$

10. 10

11. regular, decagon

12. $\overline{AN} \cong \overline{NT} \cong \overline{TJ} \cong \overline{JK} \cong \overline{KF} \cong \overline{FM} \cong \overline{MB} \cong \overline{BS} \cong \overline{SX} \cong \overline{AX}$

The Sum of All Interior Angles (page 79)

1. 180° · (8 − 2)
   180 · 6 = 1080°
   sum of interior angles = 1080°

2. 180° · (5 − 2)
   180 · 3 = 540°
   sum of interior angles = 540°

3. 180° · (8 − 2)
   180 · 6 = 1080°
   sum of interior angles = 1080°

4. 180° · (6 − 2)
   180 · 4 = 720°
   sum of interior angles = 720°

The Sum of All Exterior Angles (page 80)

1. ∠A + ∠1 = 180°
   ∠A = 108°
   ∠1 = 72°
   72 × 5 angles = 360°

2. ∠B + ∠1 = 180°
   ∠B = 90°
   ∠1 = 90°
   90 × 4 angles = 360°

3. ∠A + ∠1 = 180°
   ∠A = 128.57°
   ∠1 = 51.43°
   51.43 × 7 angles ≈ 360°

4. ∠D + ∠5 = 180°
   ∠C + ∠6 = 180°
   ∠D = 146°
   ∠5 = 34°
   34 × 2 angles = 68°
   ∠C = 107°
   ∠6 = 73°
   73 × 4 angles = 292°
   292° + 68° = 360°

Polygon Congruence (page 81)

1. ∠X ≅ ∠S ; ∠Z ≅ ∠R ; ∠Y ≅ ∠T ; $\overline{ST} ≅ \overline{XY}$; $\overline{XZ} ≅ \overline{RS}$; $\overline{RT} ≅ \overline{YZ}$

2. ∠A ≅ ∠W ; ∠B ≅ ∠X ; ∠D ≅ ∠Y ; ∠C ≅ ∠Z ; $\overline{AB} ≅ \overline{XW}$; $\overline{BC} ≅ \overline{XZ}$; $\overline{WY} ≅ \overline{AD}$; $\overline{CD} ≅ \overline{YZ}$

More Congruent Polygons (page 82)

1. ∠PQM ≅ ∠RQM; ∠MPQ ≅ ∠MRQ; ∠PMQ ≅ ∠RMQ $\overline{MP} ≅ \overline{MR}$; $\overline{PQ} ≅ \overline{QR}$; $\overline{QM} ≅ \overline{QM}$

2. ∠F ≅ ∠R; ∠G ≅ ∠Z; ∠W ≅ ∠H; ∠J ≅ ∠T $\overline{FG} ≅ \overline{RZ}$; $\overline{JH} ≅ \overline{TW}$; $\overline{RT} ≅ \overline{FJ}$; $\overline{GH} ≅ \overline{WZ}$

3. ∠FCD ≅ ∠ACB; ∠CFD ≅ ∠ABC; ∠CDF ≅ ∠CAB $\overline{CD} ≅ \overline{CA}$; $\overline{CB} ≅ \overline{CF}$; $\overline{FD} ≅ \overline{AB}$

4. ∠M ≅ ∠S ; ∠Q ≅ ∠R ; ∠P ≅ ∠T ; ∠N ≅ ∠V $\overline{RT} ≅ \overline{PQ}$; $\overline{TS} ≅ \overline{MP}$; $\overline{SV} ≅ \overline{MN}$; $\overline{RV} ≅ \overline{NQ}$

Find Measures with Congruent Polygons (page 83)

1. m∠M = m∠R = 60°

   m∠O = m∠S = 65°

   m∠N = m∠T = 55°

   m∠M = 60°; m∠T = 55°; m∠S = 65°

2. m∠L = m∠M = 88°

   m∠J = m∠B = 122°

   m∠K = m∠S = 65°

   m∠P = m∠Y = 85°

   m∠J = 122°; m∠K = 65°; m∠Y = 85°; m∠M = 88°

Determine Congruence by Sides or Angles (page 84)

1. $\overline{JL} ≅ \overline{RS}$

   ∠JLK ≅ ∠RST

   $\overline{LK} ≅ \overline{ST}$

2. SAS

3. △JLK ≅ △RST

4. $\overline{AB} ≅ \overline{XY}$

   $\overline{AC} ≅ \overline{XZ}$

   $\overline{BC} ≅ \overline{YZ}$

5. SSS

6. △ABC ≅ △XYZ

Use SSS or SAS (page 85)

1. SSS

2. $\overline{CD} ≅ \overline{GH}$

   $\overline{DE} ≅ \overline{HJ}$

   $\overline{CE} ≅ \overline{GJ}$

3. SSS

4. $\overline{BC} ≅ \overline{DF}$

   $\overline{BF} ≅ \overline{CD}$

   $\overline{CF} ≅ \overline{FC}$

5. SAS

6. $\overline{QP} ≅ \overline{NP}$

   ∠MPQ ≅ ∠MPN

   $\overline{MP} ≅ \overline{PM}$

7. SAS

8. $\overline{AC} ≅ \overline{CD}$

   ∠ACB ≅ ∠ECD

   $\overline{BC} ≅ \overline{EC}$

9. SSS

10. $\overline{PQ} ≅ \overline{FM}$

    $\overline{PR} ≅ \overline{FN}$

    $\overline{QR} ≅ \overline{MN}$

11. SSS

12. $\overline{SW} ≅ \overline{WV}$

    $\overline{ST} ≅ \overline{TV}$

    $\overline{WT} ≅ \overline{TW}$

13. SAS

14. $\overline{GH} ≅ \overline{AC}$

    ∠GHJ ≅ ∠CAB

    $\overline{HJ} ≅ \overline{AB}$

15. SSS

16. $\overline{DE} ≅ \overline{QV}$

    $\overline{EF} ≅ \overline{VW}$

    $\overline{DF} ≅ \overline{QW}$

17. SAS

18. $\overline{TW} ≅ \overline{PQ}$

    ∠TWV ≅ ∠PQR

    $\overline{WV} ≅ \overline{QR}$

Similarity (page 86)

1. $\dfrac{AB}{FG}=\dfrac{4}{8}=\dfrac{1}{2}$ $\dfrac{BC}{GH}=\dfrac{4}{8}=\dfrac{1}{2}$ $\dfrac{CD}{HI}=\dfrac{4}{8}=\dfrac{1}{2}$ $\dfrac{AD}{FI}=\dfrac{4}{8}=\dfrac{1}{2}$

2. yes

3. polygon ABCD ~ polygon FGHI

4. $\dfrac{AB}{FD}=\dfrac{6}{12}=\dfrac{1}{2}$ $\dfrac{CB}{DE}=\dfrac{4}{8}=\dfrac{1}{2}$ $\dfrac{AC}{EF}=\dfrac{6}{13}$

5. no

6. triangle ABC is not similar to triangle DEF

Check for Similar Polygons (page 87)

1. $\dfrac{DE}{AB}=\dfrac{6}{9}=\dfrac{2}{3}$ $\dfrac{DF}{AC}=\dfrac{10}{15}=\dfrac{2}{3}$ $\dfrac{EF}{BC}=\dfrac{8}{12}=\dfrac{2}{3}$

2. yes

3. $\triangle DEF \sim \triangle ABC$

4. $\dfrac{AB}{FG}=\dfrac{4}{6}=\dfrac{2}{3}$ $\dfrac{BC}{GH}=\dfrac{6}{9}=\dfrac{2}{3}$ $\dfrac{CD}{HI}=\dfrac{6}{9}=\dfrac{2}{3}$

5. yes

6. polygon ABCD ~ polygon FGHI

7. $\dfrac{RS}{MP}=\dfrac{3}{9}=\dfrac{1}{3}$ $\dfrac{ST}{MN}=\dfrac{2}{6}=\dfrac{1}{3}$ $\dfrac{RT}{NP}=\dfrac{3}{9}=\dfrac{1}{3}$

8. yes

9. $\triangle RST \sim \triangle MNP$

Similar or Congruent? (page 88)

1. neither; angles and sides are not congruent

2. congruent; angles are congruent

3. congruent; angles and sides are congruent

4. congruent; sides are congruent

5. similar; angles are the same degrees

6. neither; angles and sides are not congruent

7. similar; angles are the same degrees

8. similar; angles are congruent

9. congruent; angles and sides are congruent

Perimeter of Polygons (page 89)

1. 5 + 5 + 9 + 9 = 28
28 units

2. 16 × 4 = 64
64 units

3. 6 + 6 + 10 + 10 = 32
32 units

4. 8 + 7 + 5 + 8 + 15 = 43
43 units

5. 11 × 5 = 55
55 units

6. 5 + 5 + 4 + 4 + 4 = 22
22 units

7. 8 + 6 + 5 + 7 = 26
26 units

8. 6 + 3 + 4 + 10 + 4 = 27
27 units

9. (3 × 10) + (9 × 2) = 48
48 units

Perimeter with Mixed Measures (page 90)

1.
```
  2 ft   7 in
  2 ft   7 in
  3 ft   5 in
+ 3 ft  11 in
 10 ft  30 in   =   12 ft  6 in
```
The perimeter is 12 feet 6 inches.

2.
```
 11 yd 1 ft
  9 yd 2 ft
+ 8 yd
 28 yd 3 ft   =   29 yds
```
The perimeter is 29 yards.

3.
```
  2 ft  10 in
  3 ft   2in
  1 ft   3 in
+ 2 ft   8 in
  8 ft  23 in   =   9 ft  11 in
```
The perimeter is 9 feet 11 inches.

Area of Squares and Rectangles (page 91)

1. A = lw
A = 11 x 5
A = 55 square units

2. A = lw
A = 4 × 6
A = 24 square units

3. $A = s^2$
A = 12² or 12 × 12
A = 144 square units

4. $A = s^2$
A = 9² or 9 × 9
A = 81 square units

Area of Parallelograms and Rhombi (page 92)

1. A = bh
A = 10 × 10
A = 100 square units

2. A = bh
A = 7 × 5
A = 35 square units

3. A = bh
A = 12 × 6
A = 72 square units

4. A = bh
A = 9 × 17
A = 153 square units

Area of Trapezoids (page 93)

1. A = $\frac{1}{2}$(5 + 10)9

A = $\frac{1}{2}$(15)9

A = $\frac{1}{2}$(135)

Area of trapezoid: 67.5 square units

2. A = $\frac{1}{2}$(9 + 5)6

A = $\frac{1}{2}$(14)6

A = $\frac{1}{2}$(84)

Area of trapezoid: 42 square units

3. A = $\frac{1}{2}$(5 + 7)4

A = $\frac{1}{2}$(12)4

A = $\frac{1}{2}$(48)

Area of trapezoid: 24 square units

Area of Triangles (page 94)

1. A = $\frac{1}{2}$bh

A = $\frac{1}{2}$(11 × 7)

A = $\frac{1}{2}$(77)

A = 38.5 square units

2. A = $\frac{1}{2}$bh

A = $\frac{1}{2}$(18 × 8)

A = $\frac{1}{2}$(144)

A = 72 square units

3. A = $\frac{1}{2}$bh

A = $\frac{1}{2}$(4 × 5)

A = $\frac{1}{2}$(20)

A = 10 square units

4. A = $\frac{1}{2}$bh

A = $\frac{1}{2}$(6 × 10)

A = $\frac{1}{2}$(60)

A = 30 square units

Area of Perpendicular Polygons (page 95)

Note: Each polygon can be divided into a different set of rectangles.

1. 4 × 5 = 20
4 × 3 = 12
4 × 5 = 20

2. A = 20 + 12 + 20
A = 52 square units

3. 4 × 2 = 8
9 × 5 = 45
3 × 3 = 9

4. A = 8 + 45 + 9
A = 62 square units

Formulas for Areas (page 96)

1. C
21 square units

2. C and A
192 square units

3. C
28 square units

4. B
50 square units

5. A
55 square units

6. D
25 square units

Find Areas (page 97)

1. A = $\frac{1}{2}$bh
99 square units

2. A = bh
105 square units

3. A = $\frac{1}{2}$bh
28 square units

4. A = (lw) + (lw)
122 square units

5. A = lw
154 square units

6. A = $\frac{1}{2}$h(b$_1$ + b$_2$)
77 square units

Words for Circles (page 98)

1. concentric
2. secant
3. circle
4. diameter
5. tangent
6. arc
7. chord
8. radius
9. interior
10. semicircle
11. exterior
12. congruent
13. center

Label Circle Parts (page 99)

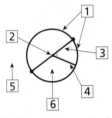

1. semicircles
2. center
3. diameter
4. radius
5. exterior
6. interior

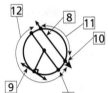

7. diameter
8. chord
9. tangent
10. secant
11. minor arc
12. major arc

Congruent and Concentric Circles (page 100)

1. congruent
   Circle C = 3 cm
   Circle D = 3 cm
2. concentric
   Circle F = 1 ft
   Circle G = 5.5 ft
3. concentric
   Circle M = 25 cm
   Circle N = 30.5 cm
4. congruent
   Circle Q = 5.5 in
   Circle R = 5.5 in

Tangent Circles (page 101)

1. internally tangent
2. Point M
3. $\overleftrightarrow{AC}$
4. externally tangent
5. Point Q
6. $\overline{LP}$
7. externally tangent
8. Point X

9. $\overleftrightarrow{EG}$
10. internally tangent
11. Point R
12. $\overleftrightarrow{BT}$

Inscribed and Circumscribed Circles (page 102)

1. inscribed
2. Points B, D, F, H, and J
3. 5 tangent points
4. circumscribed
5. Points G, H, J, K, M, and N
6. 6 tangent points
7. inscribed
8. Points S, M, B, P, O, and A
9. 6 tangent points
10. circumscribed
11. Points M, P, B, and C
12. 4 tangent points

Diameters and Radii (page 103)

1. 3.75 cm
   7.5 cm
   $\overleftrightarrow{CD}$
   $\overline{AF}$ or $\overline{CD}$
2. 6 cm
   12 cm
   $\overleftrightarrow{MO}$
   $\overline{MO}$
3. 7 cm
   14 cm
   $\overleftrightarrow{JH}$
   $\overline{JH}$ or $\overline{DH}$
4. 3.75 cm
   7.5 cm
   $\overleftrightarrow{TU}$
   $\overline{RT}$ or $\overline{TU}$
5. 5.5 cm
   11 cm
   $\overleftrightarrow{LN}$
   $\overline{BD}$ or $\overline{LN}$
6. 7 cm
   14 cm
   $\overleftrightarrow{RS}$
   $\overline{RS}$ or $\overline{TV}$

Area of a Circle (page 104)

1. 4.5 cm
$A = 3.14(4.5)^2$
$A = 3.14(20.24)$
$A = 63.585$
Area = 63.585 cm

2. 6 m
$A = 3.14(6)^2$
$A = 3.14(36)$
$A = 113.04$
Area = 113.04 cm

3. 2.5 cm
$A = 3.14(2.5)^2$
$A = 3.14(6.25)$
$A = 19.625$
Area = 19.625 cm

4. 7.5 cm
$A = 3.14(7.5)^2$
$A = 3.14(56.25)$
$A = 176.625$
Area = 176.625 cm

Arcs and Angles (page 105)

1. $\overarc{CAB}$ or $\overarc{ABC}$

2. $\overarc{CB}$ or $\overarc{CA}$

3. ∠BDC or ∠ADC

4. $\overarc{AB}$ or $\overarc{BA}$

5. $\overarc{JKM}$ or $\overarc{KJN}$ or $\overarc{KNJ}$ or $\overarc{KMJ}$ or $\overarc{NMK}$

6. $\overarc{KN}$ or $\overarc{NM}$ or $\overarc{JM}$ or $\overarc{JK}$ or $\overarc{JN}$

7. ∠JLK or ∠JLM

8. $\overarc{KM}$ or $\overarc{MK}$

9. Possible answers include: $\overarc{STV}$, $\overarc{SRV}$, $\overarc{WSR}$, $\overarc{VSR}$, (any more than 180°)

10. Possible answers include: $\overarc{ST}$, $\overarc{WV}$, $\overarc{VR}$, $\overarc{TR}$, (any less than 180°)

11. ∠RQT or ∠RQV

12. $\overarc{VT}$ or $\overarc{TV}$

13. $\overarc{FDH}$ or $\overarc{HFD}$

14. $\overarc{FH}$ or $\overarc{DH}$

15. ∠FGH or ∠DGH

16. $\overarc{FD}$ or $\overarc{DF}$

Reflections (page 106)

1. reflection   **2.** no reflection

3. reflection   **4.** no reflection

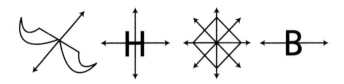

Line Symmetry (page 107)

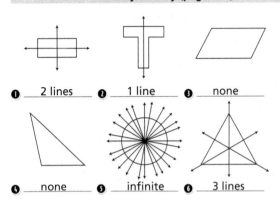

1 2 lines   **2** 1 line   **3** none

4 none   **5** infinite   **6** 3 lines

Practice with Symmetry (page 108)

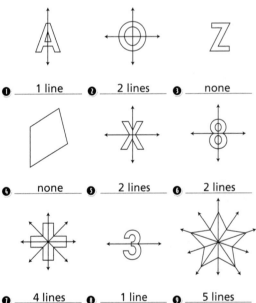

1 1 line   **2** 2 lines   **3** none

4 none   **5** 2 lines   **6** 2 lines

7 4 lines   **8** 1 line   **9** 5 lines

Rotations (page 109)

1. clockwise

2. 1/4 turn

3. outside of figure

4. counter-clockwise or clockwise

5. 1/2 turn

6. Point E

7. counter-clockwise

8. 1/4 turn

9. outside of figure

10. counter-clockwise

11. 1/4 turn

12. Point T

Rotational Symmetry (page 110)

❶ _____ yes   ❷ _____ yes

❸ _____ yes   ❹ _____ no

5. Possible answer: Regular polygons have rotational symmetry.

Translations (page 111)

1. $\overline{CD} \parallel$ and $\cong \overline{C'D'}$

   $\overline{DF} \parallel$ and $\cong \overline{D'F'}$

   $\overline{CF} \parallel$ and $\cong \overline{C'F'}$
   slides over

2. $\overline{PM} \parallel$ and $\cong \overline{P'M'}$

   $\overline{PN}$ and $\cong \overline{P'N'}$

   $\overline{MN} \parallel$ and $\cong \overline{M'N'}$
   slides over, then down

3. $\overline{XY} \parallel$ and $\cong \overline{X'Y'}$

   $\overline{XZ} \parallel$ and $\cong \overline{X'Z'}$

   $\overline{YZ} \parallel$ and $\cong \overline{Y'Z'}$
   slides over, then up

4. $\overline{AB} \parallel$ and $\cong \overline{A'B'}$

   $\overline{BC} \parallel$ and $\cong \overline{B'C'}$

   $\overline{CD} \parallel$ and $\cong \overline{C'D'}$

   $\overline{AD} \parallel$ and $\cong \overline{A'D'}$
   slides over, then up

Practice with Transformations (page 112)

1. reflection

2. Figure is reflected or reversed.

3. rotation

4. Figure is turned on outside point clockwise 1/4 turn.

5. translation

6. Figure slides over and down.

7. rotation

8. Figure made 1/4 turn clockwise on outside point.

9. translation

10. Figure slides over and down.

11. reflection and translation

12. Figure moves clockwise. First, it is reflected, then reflected again. Then it is rotated 1/2 turn clockwise. Then it is slid up.

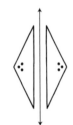

 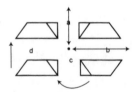